*Gerhard K. Ackermann
and Jürgen Eichler*
Holography

1807–2007 Knowledge for Generations

Each generation has its unique needs and aspirations. When Charles Wiley first opened his small printing shop in lower Manhattan in 1807, it was a generation of boundless potential searching for an identity. And we were there, helping to define a new American literary tradition. Over half a century later, in the midst of the Second Industrial Revolution, it was a generation focused on building the future. Once again, we were there, supplying the critical scientific, technical, and engineering knowledge that helped frame the world. Throughout the 20th Century, and into the new millennium, nations began to reach out beyond their own borders and a new international community was born. Wiley was there, expanding its operations around the world to enable a global exchange of ideas, opinions, and know-how.

For 200 years, Wiley has been an integral part of each generation's journey, enabling the flow of information and understanding necessary to meet their needs and fulfill their aspirations. Today, bold new technologies are changing the way we live and learn. Wiley will be there, providing you the must-have knowledge you need to imagine new worlds, new possibilities, and new opportunities.

Generations come and go, but you can always count on Wiley to provide you the knowledge you need, when and where you need it!

William J. Pesce
President and Chief Executive Officer

Peter Booth Wiley
Chairman of the Board

Gerhard K. Ackermann and Jürgen Eichler

Holography

A Practical Approach

WILEY-VCH Verlag GmbH & Co. KGaA

The Authors

Prof. Dr. rer. nat. Gerhard K. Ackermann
President, retired
Technische Fachhochschule Berlin
University of Applied Sciences
Seestr. 64
13347 Berlin
Germany
nnamrecka@arcor.de

Prof. Dr. rer. nat. Jürgen Eichler
Technische Fachhochschule Berlin
University of Applied Sciences
Seestr. 64
13347 Berlin
Germany
juergen.eichler@tfh-berlin.de

Translation

Jan Müller, Hamburg

Illustrations

Peter Hesse, Berlin

Cover

"MATRIX 18R", 1985, silver halide emulsion on glass is one of a series of Kinetic Yantras by Fred Unterseher.
Technically the hologram can best be described as an off axis Fourier transform lens matrix, holographic optical element (H.O.E.). This particular technique produces a white light viewable hologram of pure dimensional light alone, and allows for a greater degree of spontaneity in the process of making a hologram. The holographic imagery appears as a kinetic form of pure light, instead of reflected light from a given object.
e-mail: fred.unterseher@gmail.com

All books published by Wiley-VCH are carefully produced. Nevertheless, authors, editors, and publisher do not warrant the information contained in these books, including this book, to be free of errors. Readers are advised to keep in mind that statements, data, illustrations, procedural details or other items may inadvertently be inaccurate.

Library of Congress Card No.:
applied for

British Library Cataloguing-in-Publication Data
A catalogue record for this book is available from the British Library.

Bibliographic information published by the Deutsche Nationalbibliothek
Die Deutsche Nationalbibliothek lists this publication in the Deutsche Nationalbibliografie; detailed bibliographic data are available in the Internet at <http://dnb.d-nb.de>.

© 2007 WILEY-VCH Verlag GmbH & Co. KGaA, Weinheim

All rights reserved (including those of translation into other languages). No part of this book may be reproduced in any form – by photoprinting, microfilm, or any other means – nor transmitted or translated into a machine language without written permission from the publishers. Registered names, trademarks, etc. used in this book, even when not specifically marked as such, are not to be considered unprotected by law.

Typesetting   Uwe Krieg, Berlin
Printing   betz-druck GmbH, Darmstadt
Binding   Litges & Dopf GmbH, Heppenheim
Wiley Bicentennial Logo   Richard J. Pacifico

Printed in the Federal Republic of Germany
Printed on acid-free paper

ISBN: 978-3-527-40663-0

Contents

Preface *XVII*

Part 1 Fundamentals of Holography *1*

1 Introduction *3*
1.1 Photography and Holography *3*
1.1.1 Object Wave *3*
1.1.2 Photography *4*
1.1.3 Holography *4*
1.2 Interference and Diffraction *6*
1.2.1 Interference During Recording *6*
1.2.2 Diffraction During Reconstruction *7*
1.3 History of Holography *7*
Problems *7*

2 General View of Holography *9*
2.1 Interference of Light Waves *9*
2.1.1 Wave *9*
2.1.2 Interference *11*
2.1.3 Visibility *13*
2.1.4 Influence of Polarization *13*
2.2 Holographic Recording and Reconstruction *13*
2.2.1 Recording *14*
2.2.2 Reconstruction *14*
2.3 Mathematical Approach *16*
2.3.1 Object and Reference Wave *16*
2.3.2 Recording *16*
2.3.3 Gratings *18*
2.3.4 Reconstruction *18*

Holography: A Practical Approach. Gerhard K. Ackermann and Jürgen Eichler
Copyright © 2007 WILEY-VCH Verlag GmbH & Co. KGaA, Weinheim
ISBN: 978-3-527-40663-0

2.4	Conjugated Image 19
2.4.1	Conjugated Object Wave 19
2.4.2	Position of the Conjugated Image 21
2.4.3	Reversal of the Reconstruction Wave 21
2.5	Spatial Frequencies 22
2.6	Diffraction Grating and Fresnel Lens 23
2.6.1	Diffraction Grating 24
2.6.2	Fresnel Zone Lens 25
	Problems 27

3	**Fundamental Imaging Techniques in Holography** 29
3.1	In-Line Hologram (Gabor) 29
3.2	Off-Axis Hologram (Leith–Upatnieks) 31
3.3	Fourier Hologram (Lensless) 32
3.4	Fraunhofer Hologram 34
3.5	Reflection Hologram (Denisyuk) 35
	Problems 36

4	**Holograms of Holographic Images** 39
4.1	Image-Plane Hologram 39
4.2	Transmission and Reflection Hologram in Two Steps 39
4.3	Rainbow Hologram 42
4.4	Double-Sided Hologram 44
4.5	Fourier Hologram 46
4.5.1	Principle 46
4.5.2	Calculation 46
	Problems 48

5	**Optical Properties of Holographic Images** 49
5.1	Hologram of an Object Point 49
5.1.1	Image Equations 49
5.1.2	Magnification 50
5.1.3	Angular Magnification 51
5.1.4	Longitudinal Magnification 51
5.1.5	Image Aberrations 52
5.2	Properties of the Light Source 52
5.2.1	Spectral Bandwidth 53
5.2.2	Image-Plane Holograms 53
5.3	Image Luminance 54
5.3.1	Without Pupil 54
5.3.2	With Pupil 55
5.3.3	Image-Plane Holograms 56

5.4	Speckles 56	
5.4.1	Diffuser 56	
5.4.2	Resolution 57	
5.4.3	Incoherent Illumination 57	
5.4.4	Further Techniques 57	
5.5	Resolution 58	
	Problems 58	
6	**Types of Holograms** 59	
6.1	Introduction 59	
6.1.1	Transmission and Reflection Holograms 59	
6.1.2	Thick and Thin Holograms 59	
6.2	Thin Holograms 60	
6.2.1	Thin Amplitude Holograms 60	
6.2.2	Thin Phase Holograms 61	
6.3	Volume Holograms 64	
6.3.1	Theory of Coupled Waves 64	
6.3.2	Phase Holograms 67	
6.3.3	Amplitude Holograms 68	
6.3.4	Comparison of Diffraction Efficiency 69	
6.3.5	Distinction Criteria for Holograms 70	
	Problems 71	
Part 2	**Basic Experiments** 73	
7	**Optical Systems and Lasers for Holography** 75	
7.1	Coherence and Interferometers 75	
7.1.1	Coherence 75	
7.1.2	Spatial Coherence 75	
7.1.3	Temporal Coherence 76	
7.2	Modes and Coherence 78	
7.2.1	Gaussian Beam 78	
7.2.2	Longitudinal Modes 79	
7.2.3	Coherence Length 80	
7.2.4	Etalon 81	
7.3	Gas Lasers for Holography 82	
7.3.1	He–Ne Laser 82	
7.3.2	Ion Laser 83	
7.3.3	He–Cd Laser 84	
7.4	Solid-State Lasers for Holography 84	
7.4.1	Ruby Laser 84	

7.4.2	Nd:YAG Laser	86
7.5	Lenses and Spatial Filters	86
7.5.1	Gaussian Beam	86
7.5.2	Focusing	87
7.5.3	Geometrical Optics	88
7.5.4	Spatial Filters	89
7.5.5	Beam Expansion	90
7.6	Polarizers and Beam Splitters	91
7.6.1	Polarization	91
7.6.2	Dichroitic Filters	91
7.6.3	Polarization by Reflection	91
7.6.4	Polarization Prisms	92
7.6.5	Thin Film Polarizers	93
7.6.6	$\lambda/4$- and $\lambda/2$-plates	93
7.6.7	Beam Splitter	94
7.6.8	Metal Mirrors	95
7.6.9	Dielectric Multilayer Mirrors	95
7.6.10	Nonreflective Coating	95
7.6.11	Laser Mirrors	96
7.7	Vibration Isolation	96
7.7.1	Isolators	97
7.7.2	Table Tops	99
7.7.3	Vibration Isolated Table	99
7.8	Optical Fibers and Diode Lasers	101
7.8.1	Monomode Fibers	101
7.8.2	Diode Lasers	102
	Problems	102
8	**Basic Experiments in the Holographic Laboratory**	**105**
8.1	Polarization and Brewster Angle	105
8.1.1	Experiment 1: Analyzer and Polarizer	105
8.1.2	Experiment 2: Rotation of the Polarization Plane	105
8.1.3	Experiment 3: Brewster Angle	106
8.1.4	Experiment 4: Variable Beam Splitter	107
8.2	Experiments with Lenses	107
8.2.1	Experiment 5: Measuring Focal Length	107
8.2.2	Experiment 6: Adjusting Lenses	108
8.2.3	Experiment 7: Adjusting a Spatial Filter	108
8.3	Experiments on Diffraction and Interference	110
8.3.1	Experiment 8: Diffraction at an Edge or Slit	110
8.3.2	Experiment 9a: Interferences at a Glass Plate	111
8.3.3	Experiment 9b: Newton Rings and "Index-Matching"	111

8.3.4	Experiment 10: Diffraction of a Grating	*112*
8.3.5	Experiment 11: Divergence of the Laser Beam	*113*
8.3.6	Experiment 12: Optical Filtering	*114*
8.3.7	Experiment 13: Granulation of Laser Radiation	*115*
8.4	Measurements with Interferometers	*116*
8.4.1	Experiment 14: Setup of a Michelson Interferometer	*116*
8.4.2	Experiment 15: Measuring Coherence Length	*117*
8.4.3	Experiment 16: Investigation of Stability	*118*
8.5	Production of Gratings and Simple Holograms	*118*
8.5.1	Experiment 17: Production of Diffraction Gratings	*118*
8.5.2	Exposure and Developing	*120*
8.5.3	Experiment 18: White Light Hologram	*120*
8.5.4	Experiment 19: Transmission Hologram	*121*
8.6	Experiments in the Darkroom	*121*
8.6.1	Experiment 20: Developing	*121*
8.6.2	Experiment 21: Solving Bleach Bath	*122*
8.6.3	Experiment 22: Rehalogenizing Bleach Bath	*122*
	Problems	*123*
9	**Experimental Setups for Single-Beam Holography**	*125*
9.1	Setups for Reflection Holograms	*125*
9.1.1	Experimental Setups	*125*
9.1.2	Index Matching of Holographic Films	*126*
9.1.3	Setups without Index Matching	*127*
9.1.4	Intensity Loss at Brewster-Angle Setting	*127*
9.1.5	Vacuum Film Support	*128*
9.1.6	Simple Single-Beam Setups	*128*
9.1.7	Holographic Table	*129*
9.1.8	Visibility	*129*
9.2	Setups for Transmission Holograms	*129*
9.3	Experiments for Image Reconstruction	*131*
9.3.1	Reconstruction Angle	*131*
9.3.2	Reconstruction Light Source for Transmission Holograms	*131*
9.3.3	Reconstruction of the Real Image	*132*
9.3.4	Reconstruction of Reflection Holograms	*133*
9.3.5	Wavelength Shift	*133*
9.3.6	Reconstruction of the Real Image	*134*
9.4	Trouble Shooting	*134*
	Problems	*135*

Part 3 Advanced Experiments and Materials *137*

10 Experimental Setups for Split-Beam Holography *139*
10.1 Setups for Transmission Holograms *139*
10.1.1 Experimental Setup *139*
10.1.2 Vibration *140*
10.1.3 Object *140*
10.1.4 Avoidance of Scattered Light *141*
10.1.5 Index Matching *141*
10.1.6 Visibility *141*
10.1.7 Reconstruction of the Object Waves *142*
10.1.8 Reconstruction of the Reference Wave *142*
10.2 Setups for Reflection Holograms *143*
10.2.1 Object *144*
10.2.2 Reconstruction *144*
Problems *145*

11 Experimental Setups for Holograms of Holographic Images *147*
11.1 Master Hologram (H1) *147*
11.1.1 Size and Position of the Object *147*
11.1.2 Object Distance and Position of the Reference Wave *148*
11.1.3 Preparation of the Reference Wave *148*
11.1.4 Duplicating Methods *149*
11.2 White Light Reflection Hologram (H2) *150*
11.2.1 Single-Beam Method *150*
11.2.2 Two-Beam Method *150*
11.2.3 Image Aberrations *151*
11.3 White Light Transmission Hologram (H2) *152*
11.3.1 Image Plane Hologram *152*
11.3.2 Rainbow Hologram *153*
11.3.3 Reconstruction *154*
11.3.4 Calculation of a Rainbow Hologram *155*
11.3.5 Optical Basics *155*
11.3.6 Calculation Example *156*
11.3.7 Designing a Rainbow Hologram *156*
Problems *157*

12 Other Methods in Holography *159*
12.1 Shadow Hologram *159*
12.2 Single-Beam Rainbow Hologram *160*
12.3 Multiple Exposure *161*
12.4 Multiplex Holograms *162*

12.5	360° Holography	*163*
12.6	Color Holography	*163*
12.6.1	Multilaser Techniques for Transmission Holograms	*164*
12.6.2	Multilaser Techniques for Reflection Holograms	*165*
12.6.3	Color Holograms Using Rainbow Technique	*166*
12.6.4	Achromatic Images	*167*
	Problems	*167*

13   Properties of Holographic Emulsions   *169*
13.1   Transmission and Phase Curves   *169*
13.1.1   Optical Density   *169*
13.1.2   Modulation   *171*
13.1.3   Bleaching   *171*
13.2   Resolution and Diffraction Efficiency   *172*
13.2.1   Visibility Transfer Function   *172*
13.2.2   Recording   *173*
13.2.3   Efficiency   *173*
13.3   Noise of Emulsion Layers   *174*
13.3.1   Fourier Analysis   *175*
13.3.2   Measurement Procedure   *175*
13.4   Nonlinear Effects   *176*
13.4.1   Influence of Harmonics   *176*
13.4.2   Thick Holograms   *177*
    Problems   *177*

14   Recording Media for Holograms   *179*
14.1   Silver Halide Emulsions   *179*
14.1.1   Working Principle   *180*
14.1.2   Resolution   *180*
14.1.3   Spectral Resolution   *181*
14.1.4   H&D Curves   *182*
14.1.5   Diffraction Efficiency   *183*
14.1.6   Scattered Light   *183*
14.2   Exposure, Developing, and Bleaching   *183*
14.2.1   Exposure   *184*
14.2.2   Phase Holograms   *185*
14.2.3   Optical Density   *186*
14.2.4   Phase Holograms by Bleaching   *186*
14.2.5   Shrinkage of the Emulsions   *188*
14.2.6   Pseudocolors, Preswelling   *188*
14.2.7   Index Matching   *189*

14.2.8　　Developer *189*
14.2.9　　Bleaching *190*
14.3　　Dichromate Gelatin *192*
14.3.1　　Working Principle *192*
14.3.2　　Preparation of Dichromate Gelatin (DCG) Holographic Plates *192*
14.3.3　　Properties of DCG Holographic Plates *193*
14.3.4　　Exposure and Developing *193*
14.4　　Photothermoplastic Films *194*
14.4.1　　Structure of the Layers *194*
14.4.2　　Optical Properties *196*
14.5　　Photoresists *196*
14.6　　Other Recording Media *197*
14.6.1　　Photopolymers *197*
14.6.2　　Photochromic Material *198*
14.6.3　　Photorefractive Crystals *198*
　　　　　Problems *198*

Part 4　　Application of Holography *201*

15　　Holographic Interferometry *203*
15.1　　Double-Exposure Interferometry *203*
15.1.1　　Principle *203*
15.1.2　　Theory *204*
15.1.3　　Practical Realization *205*
15.1.4　　Sandwich Method *205*
15.2　　Real-Time Interferometry *206*
15.2.1　　Principle *206*
15.2.2　　Phase Difference Between o and o' *207*
15.2.3　　Intensity of the Interferograms *207*
15.2.4　　Visibility *208*
15.2.5　　Practical Realization *209*
15.2.6　　Thermoplast Film *209*
15.3　　Fundamental Equation of Holographic Interferometry *209*
15.4　　The Holo Diagram *211*
15.5　　Time Average Interferometry *213*
15.5.1　　Theory *213*
15.5.2　　Practical Realization *214*
15.6　　Speckle Interferometry *215*
　　　　　Problems *216*

16 Holographic Optical Elements 217
16.1 Lenses, Mirrors, and Gratings 217
16.1.1 Lenses and Mirrors 217
16.1.2 Focal Length 218
16.1.3 Gratings 221
16.1.4 Beam Splitters 222
16.2 Computer-Generated Holograms 223
16.2.1 Complex HOEs 223
16.2.2 Calculated HOEs 223
16.3 Electronic Holography 224
16.3.1 Angle of Diffraction in Electronic Holography 225
16.3.2 Holographic Electronic Display 225
Problems 226

17 Security and Packing 229
17.1 Embossed Holograms 229
17.1.1 Production of the Master (Figs. 17.1 and 17.2) 229
17.1.2 Production of the Shim (Fig. 17.3) 233
17.1.3 Embossing of Holograms (Fig. 17.4) 233
17.1.4 Hot Stamping (Fig. 17.5) 233
17.1.5 Properties of Embossed Holograms 233
17.1.6 Dot Matrix Hologram 234
17.1.7 Applications 234
17.2 Holographic Security Devices in Industry (Counterfeiting) 234
17.2.1 Counterfeiting Methods 235
17.2.2 Countermeasures 236
Problems 236

18 Holography and Information Technology 237
18.1 Pattern Recognition 237
18.1.1 Associative Storage 237
18.1.2 Pattern Recognition 237
18.1.3 Image Processing 238
18.2 Neuro Computer 238
18.2.1 Recognition of Information 239
18.2.2 Phase Conjugated Mirrors 240
18.3 Digital Holographic Memories 240
18.3.1 Stack Organized Memories 240
18.3.2 Stack Organized DVD 241
18.3.3 Microholographic DVD 242
Problems 242

19 Holography and Communication 245
19.1 Holographic Diffuser Display Screen 245
19.1.1 Lambertian Diffuser 245
19.1.2 Holographic Diffuser Screen [46,47] 246
19.2 Holographic Display [47] 248
19.3 Holographic TV and Movies 249
19.3.1 Holographic TV 250
19.3.2 Alternative Methods 251
19.3.3 Holographic Movie 251
19.3.4 State-of-the-Art 252
Problems 252

20 Holography – Novel Art Medium 253
20.1 Artistic Holographic Works 253
20.1.1 Regarding the Critics on the Medium Holography 253
20.1.2 Examples of Art and Holography 254
20.2 Portrait Holography 256
20.2.1 Lasers for Pulsed Portrait Holography 256
20.2.2 Master (H1) and Reflection Copy (H2) 256
20.2.3 Eye Safety Calculations 258
20.2.4 Multiplexing Method 261
Problems 262

21 Holography in Technology and Architecture 265
21.1 Holography in Solar Energy 265
21.1.1 Photovoltaic Concentration 265
21.2 Holography, Daylighting, and IR Blocking 268
21.2.1 Daylighting 268
21.2.2 Thermal Blocking 269
21.2.3 Other Applications 270
21.3 Holography in Architecture 271
21.3.1 Documentation and Visualization 271
21.3.2 Embossed Holography 271
21.3.3 Holography on Walls and Floors 271
21.3.4 HOE in Architectural Structures 272
21.4 Detection of Particles 272
21.4.1 Recording 272
21.4.2 Reconstruction 273
Problems 274

Appendix

A   **Wave Functions and Complex Numbers** *275*
A.1   Complex Numbers *275*
A.2   Wave Functions, Sine and Cosine Waves *276*
A.3   Wave Functions, Exponential Representation *278*

B   **Bragg Diffraction** *281*

C   **Fourier Transform and Fourier Hologram** *285*
C.1   Fourier Series *285*
C.2   Exponential Form of Fourier Series *288*
C.3   Fourier Transform *288*
C.4   Convolution and Correlation *289*

Solutions to the Problems *293*

References *311*

Index *315*

Preface

More than 10 years ago the authors published a first book on holography. Since then holography evolved in many areas. The new possibilities of holography, the free design of colors and forms within a hologram inspired many artists to impressive works and installations. In addition making portraits is one of the professional areas using pulse holography.

The technical application in the field of counterfeiting and in other security devices is of a high standard. Holograms are found in everyday life: credit cards, bills, visa, logos and trademarks are secured against counterfeiting by holograms. Holography penetrates technology in many areas like nondestructive testing, holographic optical elements (HOE), optoelectronic devices, holographic storages, and digital displays.

In order to incorporate the new developments and changes in holography the authors published this book on Practical Holography. In the field of holographic material many new companies are on the market and well known have given up. Computer generated holograms are used widely in the scientific community because of the high diffraction efficiency. The book takes into consideration these new developments.

The textbook is based on laboratory courses, which were offered since two decades at our University. It is designed for students and newcomers as well as for all professionals in holography. On over 300 pages it gives all necessary information to do and to understand holography. It contains more than 100 figures and more than 100 problems including the solutions. Some mathematical more complex details are handled in the three appendices.

The 60th anniversary of holography in this year is the best motivation to publish a book on this fascinating subject. We hope, that many students and interested people will enjoy reading this book. It is made to assist in first steps in holography as well as in more advanced applications.

We are very indebted to the publishing staff of the Wiley-VCH company.

We dedicate this book to Ursula , Evelyn and Sascha's family.

June 2007 *G. Ackermann and J. Eichler*

Holography: A Practical Approach. Gerhard K. Ackermann and Jürgen Eichler
Copyright © 2007 WILEY-VCH Verlag GmbH & Co. KGaA, Weinheim
ISBN: 978-3-527-40663-0

Part 1   Fundamentals of Holography

1
Introduction

With its many applications holography is one of the most interesting developments in modern optics. Its scientific importance is emphasized by awarding the 1971 Nobel prize to its inventor Denis Gabor. The term "holography" is a compound of the Greek words "holos = complete" and "graphein = to write." It denotes a procedure for three-dimensional recording and displaying of images and information without the use of lenses. Therefore holography opens up completely new possibilities in science, engineering, graphics and arts. Fields of applications are interferometric measurement techniques, image processing, holographic optical elements and memories as well as art holograms.

1.1
Photography and Holography

1.1.1
Object Wave

To see an object it has to be illuminated. In doing so light is scattered and a so-called object wave is created. This wave contains the complete optical information of the object. The light wave is characterized by two parameters: the *amplitude*, which describes the brightness, and the *phase*, which contains the shape of the object. In Fig. 1.1 two waves of different objects are shown which have the same amplitudes but different phases. The objects have the

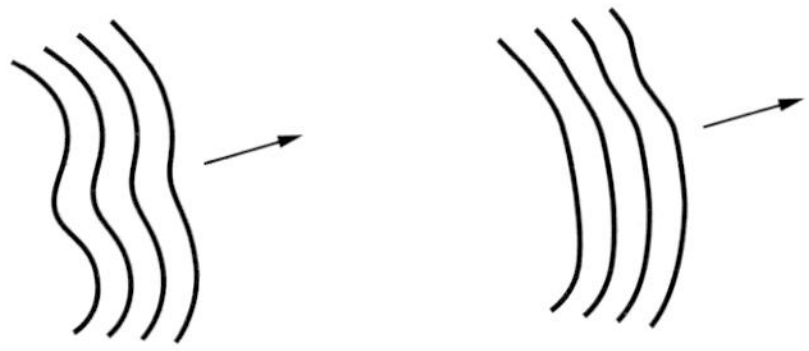

Fig. 1.1 Illustration of two light waves with same amplitudes but different phases.

Holography: A Practical Approach. Gerhard K. Ackermann and Jürgen Eichler
Copyright © 2007 WILEY-VCH Verlag GmbH & Co. KGaA, Weinheim
ISBN: 978-3-527-40663-0

same brightness but a different shape. For most holograms the color of the objects is not important, so the first chapters only deal with light waves of one wavelength. This changes for color holography which uses several wavelengths.

1.1.2
Photography

During the process of vision an object is imaged by the eye lens onto the retina. The optical path in a camera is similar: the objective creates an image on the film. For observation or to photograph an object it has to be illuminated. The scattered light, i.e., the object wave, carries the information of the object. The light wave can be made visible in a plane of the optical path, for example using a screen. The object wave appears as a very complex light field (Fig. 1.2) which results from the superposition of all waves emerging from the individual object points. If this light field could be recorded on a screen and displayed again, an observer (or a camera) would see an image that is not discriminable from the object [27].

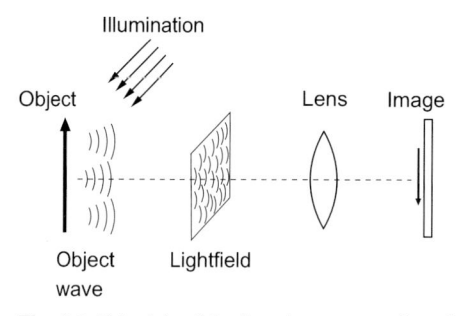

Fig. 1.2 Principle of the imaging process by a lens (camera or eye).

If there is a photographic film at the position of the screen, the object wave will cause a darkening distribution during the following processing of the film. But only the light intensity is recorded; all information of the phase in the plane of the screen is lost. This loss of phase also happens if the object is imaged onto a film by a lens. Therefore the object wave can never be completely restored from a normal photographic image. A two-dimensional image is the result.

1.1.3
Holography

Holography uses the properties interference and diffraction of light which make it possible to reconstruct the object wave completely. To be able to see these effects coherent laser light has to be used. "Coherence" means that the light wave is constant and contiguous. The laser on one hand illuminates the object and the scattered light hits the photographic film (object wave) (Fig. 1.3a). On the other hand, the film is illuminated directly with the same laser (reference wave). The object and the reference waves interfere with each other on the holographic film. This generates interference fringes in the holographic layer as are shown as a largely magnified image in Fig. 1.4. The distance of the fringes is in the region of µm which is in the order of magnitude of the light wavelength. The information of the object wave is contained in the modulation of the brightness of the fringes and in the distance of the fringes.

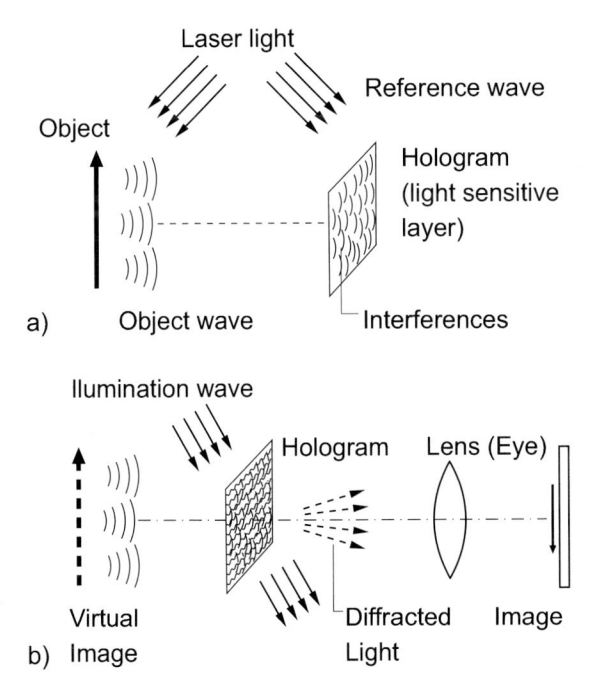

Fig. 1.3 Principle of two-stage imaging with holography: (a) recording of a hologram and (b) reconstruction of the object wave.

The photographic film is exposed and developed resulting in the hologram. The first step in holography, the *recording*, is made. The second step, the *reconstruction* or *display* of the object wave, is shown in Fig. 1.3b. After developing the film the hologram is illuminated with a light wave that should resemble the reference wave as best as possible. This reconstruction wave is diffracted by the interference pattern of the hologram generating the object wave. An observer looking at the hologram will see a three-dimensional image of the object.

Fig. 1.4 Section of a microscopic image of a hologram.

1.2
Interference and Diffraction

1.2.1
Interference During Recording

Light is an electromagnetic wave ranging from 0.4 to 0.7 µm. In the following the superposition of two constants, i.e., coherent light waves, is described. This process, known as "interference," is responsible for the recording of holograms.

A general description of the waves emerging from the object is complicated. Therefore for simplification a plane object wave is considered. The object in this case is a single point at a large distance. According to Fig. 1.5a a plane object wave and a plane reference wave impinge on the photographic layer. The superposition of the waves creates equally spaced interference fringes, i.e., parallel bright and dark areas. Dark areas occur when the waves cancel out each other by superposition of a maximum and a minimum. Bright areas occur when maxima (or minima) of the waves are superimposed. After exposing and developing the photographic layer a grating is created where exposed areas appear dark.

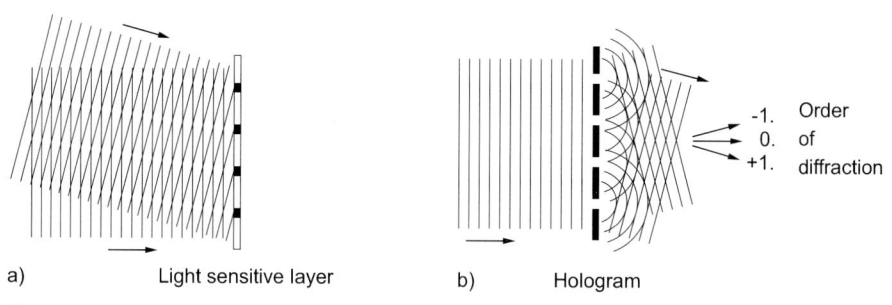

Fig. 1.5 Hologram with a plane object wave: (a) recording of the hologram (fabrication of a diffraction grating) and (b) reconstruction of the object wave (diffraction by the grating) [27].

1.2.2
Diffraction During Reconstruction

The image is displayed by illuminating the grating with a wave that closely resembles the reference wave (Fig. 1.5b). According to Huygens' principle each point of the grating sends out a spherical elementary wave. They are shown in Fig. 1.5b for the center of the bright fringes. The superposition of the elementary waves can be shown by their envelope. Plane waves are created which represent the 0th, 1st, and −1st diffraction orders [1]. (Higher order of diffraction does not occur in sinusoidal gratings.) The zeroth order is the wave passing the grating in the direction of the impinging wave. The first order represents the object wave.

Through the effect of diffraction the object wave is reconstructed; this is the principle of holography. The −1st order is often not desirable in this simple stage of holography; it is called the "conjugate object wave."

1.3
History of Holography

The physical basics of holography are optics of waves, especially interference and diffraction. The first achievements are that of C. Huygens (1629–1694), who phrased the following principle: *every point that is hit by a wave is the origin of a spherical elementary wave*. Using this statement a lot of problems of diffraction can be calculated by adding up the elementary waves. Important on the way of developing holography are also the works of T. Young (1733–1829), A.J. Fresnel (1788-1827) and J. von Fraunhofer(1877–1926). Already at the beginning of the 19th century enough knowledge was at hand to understand the principles of holography. A lot of scientist were close to the invention of this method: G. Kirchhoff (1824–1887), Lord Rayleigh (1842–1919), E. Abbe (1840–1905), G. Lippmann (1845–1921), W.L. Bragg (1890–1971), M. Wolfke and H. Boersch. But it took until 1948 when D. Gabor (1900–1979) realized the basic ideas of holography.

The origin of holography was at first connected to problems in optics of electrons. Gabor made his first groundbreaking experiments using a mercury vapor lamp. At the beginning the holographic technique was of minor importance and was forgotten for some time. It was not until the coming up of laser technology when developments in holography experienced a significant upturn. So 23 years after his experiments Gabor was awarded the Nobel prize in 1971. In 1962 the theoretical aspects of this methods were refined by E. Leith and J. Upatnieks and a year later they showed off-axis holograms. This technique marks the breakthrough for the practical application of holography.

Problems

Problem 1.1 What are the two essential elements, which describe an electromagnetic wave?

Problem 1.2 Considering the two elements mentioned in Problem 1.1, what is stored during exposure of a photograph and a hologram and what is the reason of the different results?

Problem 1.3 How is the phase of the object wave preserved during holographic exposure? Name the basic optical principles for exposure and reconstruction of a hologram.

Problem 1.4 Would it help to use coherent light in a photographic exposure in order to get a three-dimensional image?

2
General View of Holography

The basic ideas for holographic recording and reconstruction have been presented in the previous chapter in a simplified way. The next section will give a short mathematical description of holography (see [27], [3], and [66]).

2.1
Interference of Light Waves

Light is an electromagnetic wave, whereas – like within many scientific and technological applications – in holography the electrical field strength is considered only. A light wave is described by a spatial and temporal varying electrical field amplitude. The intensity I of a wave is the square of the electrical field amplitude. Within this book the object wave is abbreviated with **o**, and the reference wave with **r**. The object wave **o** and the reference wave **r** are superimposed within holographic experiments. The superposition is called interference.

The hologram represents an interference pattern that is created by the superposition of object wave **o** and reference wave **r**. The phenomenon of superposition will now be described in more detail.

2.1.1
Wave

A wave corresponds to a spatially propagating oscillation. The oscillation of the electrical field $E(t)$ at a given point, in this case the point of origin, can be described by the following equation:

$$E(t) = A\cos(2\pi f t + \varphi) = A\cos(\omega t + \varphi). \tag{2.1}$$

Here A is the amplitude of the oscillation. The parameter φ represents a phase factor which is determined for $t = 0$. For abbreviation the term "angular frequency ω" is introduced: $\omega = 2\pi f$, where f is the frequency.

The oscillation for example propagates in the z-direction; Fig. 2.1 shows a "snapshot" of the light wave. The shortest distance between two points that

Holography: A Practical Approach. Gerhard K. Ackermann and Jürgen Eichler
Copyright © 2007 WILEY-VCH Verlag GmbH & Co. KGaA, Weinheim
ISBN: 978-3-527-40663-0

oscillate with the same phase is called the wavelength λ. The time a wave with the velocity c travels a distance λ is called the period T. The reciprocal value describes the frequency $f = 1/T$. Since a point at the distance z from the point of origin starts to oscillate with a phase shift which is proportional to the time $t_0 = z/c$ the equation of the oscillation at this point looks like

$$E(t) = A\cos(\omega(t - t_0) + \varphi). \tag{2.2}$$

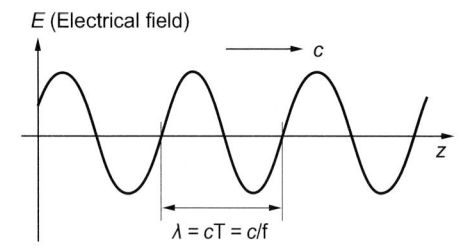

Fig. 2.1 Representation of a wave (snapshot).

From the relation $t_0 = (z/c) = z/(f \cdot \lambda)$, we obtain the equation for a plane wave, i.e., the oscillation at every point z:

$$E(z,t) = A\cos(\omega t - kz + \varphi) = A\cos(\omega t + \Phi). \tag{2.3a}$$

where $k = 2\pi/\lambda$ is called the "wave number." With this expression the plane wave is mathematically described. The phase $\Phi = \varphi - kz$ was introduced in the equation.

Generally speaking, a wave, propagating in the z-direction, has no fixed oscillation direction within the x/y plane perpendicular to z. In Fig. 2.1 the vector of the electrical field strength is oscillating in the plane of paper. Such a wave, oscillating in a fixed direction, is called linear polarized [2]. Due to technical reasons the radiation of many lasers used in holography is linear polarized.

The complex notation using Euler's relation often has advantages (see Appendix A):

$$e^{\pm i\varphi} = \cos\varphi \pm i\sin\varphi.$$

For the wave (2.3a) the complex notation is used which is denoted by bold characters:

$$\mathbf{E}(z,t) = Ae^{-i(\omega t - kz + \varphi)} = Ae^{-i(\omega t - \Phi)}. \tag{2.3b}$$

Here only the real part is important. The frequency f of a light wave is of the order of 10^{14} Hz and cannot be observed directly. In each measurement the temporal average over multiple oscillation periods is acquired; therefore the term ωt in the exponent can be left out for simplicity. The term

$$\mathbf{E} = Ae^{-i\Phi} \qquad (2.3c)$$

is called the complex amplitude.

2.1.2
Interference

After deriving the equation for a wave the superposition or interference of two light waves with the same frequency is calculated exemplary. These shall emerge from the points R and O. The polarization (alignment of the oscillation direction) of the light in Fig. 2.2 lies within the plane of the paper for optimal superposition. Both waves can be thought of as object and reference wave without restrictions to generality:

$$\mathbf{o} = oe^{-i\Phi} \qquad (2.4)$$

$$\mathbf{r} = re^{-i\Psi}. \qquad (2.5)$$

r and o are, in this case, the field amplitudes of the respective waves at the point of superposition P. The phase $\Psi = \Psi_R - 2\pi(r_1/\lambda)$ is determined by the starting phase of the wave at point R and the phase change at distance r_1. The same is valid for $\Phi = \Phi_O - 2\pi(r_2/\lambda)$.

The complex amplitudes add up at point P: $\mathbf{r} + \mathbf{o}$. The intensity I is the square of the sum of the complex amplitudes:

$$I = |\mathbf{r} + \mathbf{o}|^2 \qquad (2.6a)$$

$$I = r \cdot r^* + o \cdot o^* + o \cdot r^* + r \cdot o^*$$

$$I = r^2 + o^2 + r \cdot o \cdot \left\{ e^{-i(\Phi - \Psi)} + e^{i(\Phi - \Psi)} \right\} \qquad (2.6b)$$

$$I = r^2 + o^2 + 2 \cdot r \cdot o \cdot \cos(\Phi - \Psi).$$

If the light sources are emitting completely independently then the average of $\cos(\Phi - \Psi)$ vanishes since the phases vary statistically. This results in

$$I = r^2 + o^2 \quad \text{or}$$
$$I = I_1 + I_2. \qquad (2.7)$$

In this case the waves are called "incoherent." The intensities of both waves add up, and interference does not occur.

If the value of $\Psi_R - \Phi_O$ does not change, the waves are "coherent." According to Eq. (2.6a) locations in space exist where $\cos(\Psi - \Phi) = \pm 1$. If the field strengths oscillate in the same phase (+) this results in

$$r + o \quad \text{and} \quad I_{max} = r^2 + o^2 + 2ro. \qquad (2.8a)$$

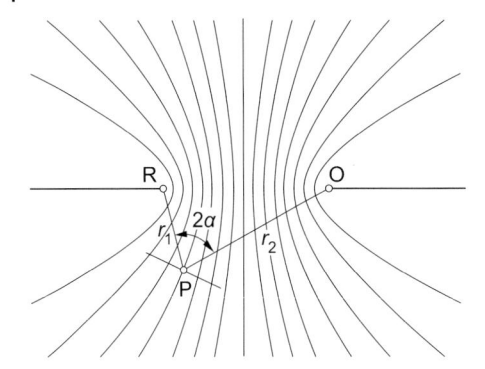

Fig. 2.2 Interference fringes or system of standing waves as generated by two coherent point light sources R and O.

If they oscillate in opposing cycles (−) the resulting superposition is

$$r - o \quad \text{and} \quad I_{min} = r^2 + o^2 - 2ro. \tag{2.8b}$$

A system of stationary waves or interference fringes forms in space. The maxima and minima are given by Eqs. (2.8a) and (2.8b). The expression $2ro\cos(\Phi - \Psi)$ in Eq. (2.6b) is called the "interference term."

The interference pattern for $\Psi_R = \Psi_O \pm N(2\pi)$ (i.e., all point sources are emitting in the same phase) is shown in Fig. 2.2. The maxima of the oscillation are given by

$$r_1 - r_2 = \pm N\lambda. \tag{2.9}$$

At the position of the maxima both waves oscillate in phase. Equation (2.9) describes a set of rotational hyperboloids. The distance of the maxima is given by

$$d = \frac{\lambda}{2\sin(\alpha)}, \tag{2.10}$$

where 2α is the angle enclosed by r_1 and r_2. For the connecting line $\overline{RO}$ is $\alpha = \pi/2$ and the distance of the maxima becomes $d = \lambda/2$.

If the interference pattern is cut perpendicular to its symmetry axis circular structures called a "Fresnel zone lens" will appear. In Section 2.6.2 a zone lens will be described as the result of the interference of a plane wave and a spherical wave. This is the case if one point, e.g. R, is shifted to infinity. If the intersection lies parallel to the connecting line $\overline{RO}$ the interference lines will form sets of hyperbolas.

2.1.3
Visibility

In holography **r** and **o** represent the reference and the object wave, respectively. During the recording of the hologram the visibility V in the interference field is given by the ratio of the two waves $I_1 = r^2$ and $I_2 = o^2$. It is defined by

$$V = \frac{I_{max} - I_{min}}{I_{max} + I_{min}}. \tag{2.11}$$

For coherent waves using Eqs. (2.8a), (2.8b), and (2.7) one gets

$$V = \frac{2 \cdot \sqrt{\frac{I_1}{I_2}}}{1 + \frac{I_1}{I_2}}. \tag{2.12a}$$

The visibility reaches a maximum of 1 at $I_1 = I_2$.

2.1.4
Influence of Polarization

For the preceding considerations concerning interference it was assumed that the polarization of the light waves is parallel. From that it follows that the maximal visibility of $V = 1$ holds for $I_1 = I_2$. If the polarization directions of two linearly polarized waves enclose an angle ψ the following equations result instead of Eqs. (2.6a) and (2.12a):

$$I = r^2 + o^2 + 2ro \cos(\Phi - \Psi) \cos \psi \tag{2.6b}$$

and

$$V = \frac{2 \cdot \sqrt{\frac{I_1}{I_2}}}{1 + \frac{I_1}{I_2}} \cos \psi. \tag{2.12b}$$

No interference occurs if the directions of polarization are perpendicular to each other; the visibility is $V = 0$. For optimal visibility object and reference wave have to polarized parallel to each other. Even by using linearly polarized radiation this cannot always be achieved in practice since light is being partly depolarized when scattered at an object [2].

2.2
Holographic Recording and Reconstruction

The difference between photography and holography lies in the ability of holography to record the intensity as well as the phase of the object wave. It may seem almost incredible that the information of a three-dimensional object, i.e., the object wave, can be recorded into a two-dimensional photographic layer. A look at the lectures about electrodynamics can help understand this principle: if the amplitude and the phase of a wave are known in one (infinite) plane, the wave field is entirely defined in space.

2.2.1
Recording

The amplitudes of the object and the reference wave on the photographic layer are given by **o** and **r**, respectively (see Fig. 2.3a). These variables describe the intensity of the electromagnetic field of the light wave which impinges on the photosensitive layer. Both waves superpose, i.e., they form **o** + **r**. In wave theory the intensity I, the brightness, is calculated as the square of the amplitude:

$$I = |\mathbf{r} + \mathbf{o}|^2 = (\mathbf{r} + \mathbf{o})(\mathbf{r} + \mathbf{o})^*. \tag{2.13a}$$

The bold letters represent complex functions which are dealt with in Section 2.1. Equation (2.13a) can be written as

$$I = |\mathbf{r}|^2 + |\mathbf{o}|^2 + \mathbf{r}\mathbf{o}^* + \mathbf{r}^*\mathbf{o}, \tag{2.13b}$$

where the star represents the complex conjugate. Especially the last term containing the object wave **o** is important for holography. The darkening of the holographic film is dependent on the intensity I. Thus the information about the object wave **o** is stored in the photographic layer. (For complex numbers see Appendix A.3.)

2.2.2
Reconstruction

The recording process of the object wave will be a lot more understandable after dealing with the following calculations for the image reconstruction. The reconstruction is performed by illuminating the hologram with the reference wave **r** (see Fig. 2.3b). For simplification we will assume that the amplitude transmission of the film material is proportional to I which is contrary to usual film processing. (Strictly speaking this assumption is of no relevance, since the effect of the diffraction pattern is the same when exchanging bright and dark fringes.) Therefore the reconstruction yields the light amplitude **u** directly behind the hologram:

$$\begin{aligned}\mathbf{u} \sim \mathbf{r} \cdot I &= \mathbf{r}(|\mathbf{o}|^2 + |\mathbf{r}|^2) + \mathbf{r}\mathbf{r}\mathbf{o}^* + |\mathbf{r}|^2 \mathbf{o} \\ &= \mathbf{u}_0 + \mathbf{u}_{-1} + \mathbf{u}_{+1}.\end{aligned} \tag{2.14}$$

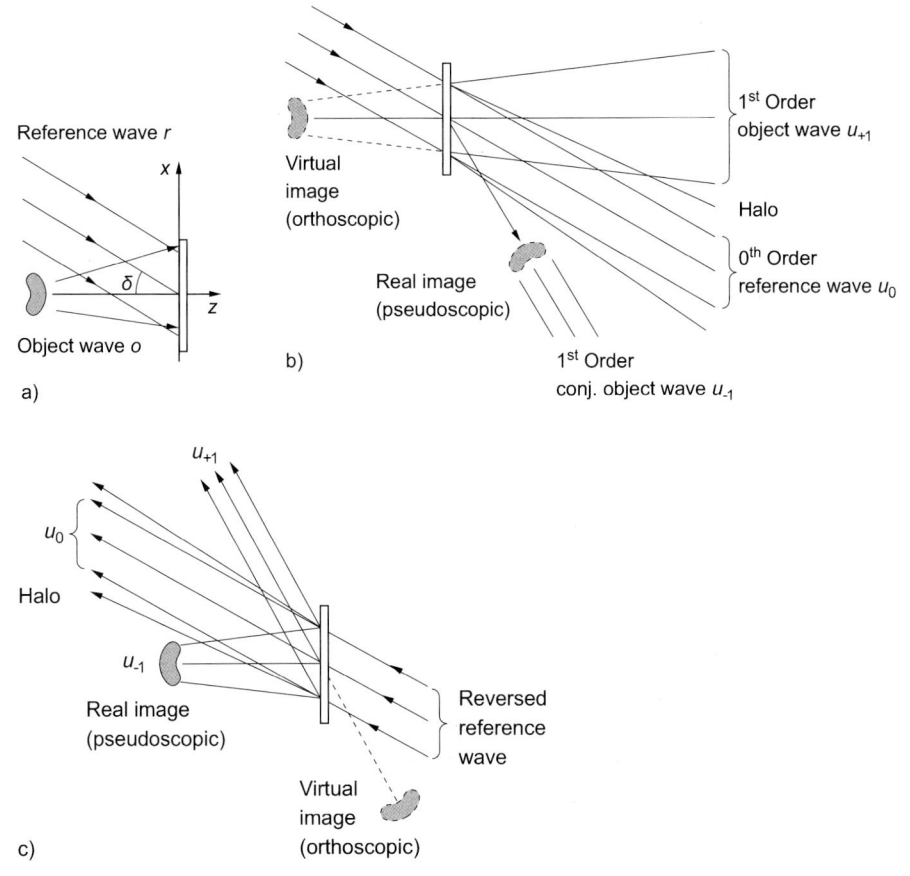

Fig. 2.3 Description of holography (off-axis hologram): (a) recording, (b) reconstruction, and (c) inversion of the reconstruction wave.

The wavefield behind the hologram is composed of three parts. The first term u_0 governs the reference wave which is weakened by the darkening of the hologram by a factor of $(|o|^2 + |r|^2)$ (zeroth diffraction order). The second term u_{-1} essentially describes the conjugate complex object wave o^*. This corresponds to the -1st diffraction order. In the last term u_{+1} the object wave itself is reconstructed with the amplitude of the reference wave $|r|^2$ being constant over the whole hologram. This proves that the object wave o can be completely reconstructed. It represents the first diffraction order.

2.3
Mathematical Approach

The principles of holography are contained in the simple Eq. (2.14) but the holographic process needs to be described more precisely. The following formulation relates to the most important method, the off-axis holography.

2.3.1
Object and Reference Wave

The complex amplitude of the object wave $\mathbf{o}(x,y)$ is a complicated function; the absolute value $|o(x,y)|$ and phase $\Phi(x,y)$ are dependent on the coordinates x and y on the photographic plate. In the following only the amplitudes in the hologram plane are described; this is sufficient to describe the whole wave field as long as the hologram is large enough. If the time dependence of the waves is not considered, Eq. (2.14) for the object wave can be written as

$$\mathbf{o}(x,y) = |o(x,y)|e^{-i\Phi}$$
$$= o(x,y)e^{-i\Phi}. \tag{2.15}$$

Complex functions are written in bold letters while the same letters in normal font are used for their absolute value. Usually the reference wave $\mathbf{r}(x,y)$ is a plane wave. The absolute value r remains constant for uniform illumination. The phase Ψ depends on the angle of incidence δ and can be calculated by (see Fig. 2.4)

$$\mathbf{r}(x,y) = re^{-i\Psi}$$
$$= re^{(i2\pi\sigma_r x)}. \tag{2.16}$$

The distance of two maxima of the reference wave in the hologram plane is given by $d_r = 1/\sigma_r$:

$$d_r = \frac{1}{\sigma_r} = \frac{\lambda}{\sin\delta}, \tag{2.17}$$

where σ_r is the so-called spatial frequency of the wave, i.e., the number of maxima per length unit.

2.3.2
Recording

The intensity in the plane of the photographic layer is therefore given by (see also Eq. (2.13b))

$$I = |\mathbf{r}(x,y) + \mathbf{o}(x,y)|^2$$
$$= |\mathbf{r}(x,y)|^2 + |\mathbf{o}(x,y)|^2 + \mathbf{r}^*(x,y)\mathbf{o}(x,y) + \mathbf{r}(x,y)\mathbf{o}^*(x,y). \tag{2.18}$$

2.3 Mathematical Approach

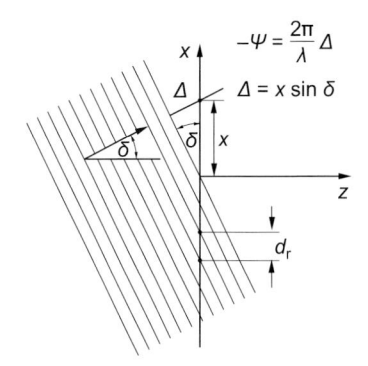

Fig. 2.4 Phase $\psi = -2\pi\Delta/\lambda$ of an angular incident wave on a hologram (sign: see Section 2.1).

The four summands can be calculated as

$$|\mathbf{r}(x,y)|^2 = r^2$$
$$|\mathbf{o}(x,y)|^2 = o^2$$
$$\mathbf{r}^*(x,y)\mathbf{o}(x,y) = ro(x,y)e^{-2\pi i\sigma_r x}e^{-i\Phi(x,y)}$$
$$\mathbf{r}(x,y)\mathbf{o}^*(x,y) = ro(x,y)e^{2\pi i\sigma_r x}e^{i\Phi(x,y)}. \tag{2.19}$$

Using the Euler formula $e^\varphi + e^{-\varphi} = 2\cos\varphi$ this can be written as

$$I(x,y) = r^2 + o^2(x,y) + 2ro(x,y)\cos[2\pi\sigma_r x + \Phi(x,y)]. \tag{2.20}$$

The equation shows that the intensity distribution in the photographic layer contains the object wave's amplitude $o(x,y)$ as well as the phase $\Phi(x,y)$. The amplitude $o(x,y)$ modulates the brightness while the phase modulates the distance of the fringes with a spatial carrier frequency σ_r (= fringes/length unit).

The properties of photographic layers are explained in more detail in Chapter 13. The transmission decreases proportional to the exposure intensity I and the exposure time τ. The transmission without any exposure is given by t_0:

$$t = t_0 + \beta\tau I = t_0 + \beta E. \tag{2.21}$$

The term $E = I\tau$ describes the energy density of the light, commonly called the "exposure." The parameter β is negative and is represented by the slope in the H&D curve (see Fig. 13.1). The amplitude transmission is then given by

$$t(x,y) = t_0 + \beta\tau r^2 \tag{2.22a}$$
$$+ \beta\tau o^2(x,y)$$
$$+ \beta\tau ro(x,y)e^{-i2\pi\sigma_r x}e^{-i\Phi(x,y)}$$
$$+ \beta\tau ro(x,y)e^{i2\pi\sigma_r x}e^{i\Phi(x,y)}.$$

2.3.3
Gratings

The transmission t of a plane object wave **o** which illuminates the photographic layer similar to the reference wave in Eq. (2.16) can be easily calculated: $\mathbf{o} = oe^{i2\pi\sigma_0 x}$. Combining Eq. (2.22a) with $\Phi = -2\pi\sigma_0 x$ and the Euler equation then gives

$$t(x) = \bar{t} + t_1 \cos(kx) \tag{2.22b}$$
$$k = 2\pi(\sigma_r + \sigma_o)$$
$$\bar{t} = t_0 + \beta\tau(r^2 + o^2(x,y))$$
$$t_1 = \beta\tau r o.$$

The amplitude transmission t of a hologram formed by two plane waves **r** and **o** is therefore a cosine-like diffraction grating. Hence the intensity transmission $T = t^2$ is proportional to a $\cos^2$-function.

2.3.4
Reconstruction

For the reconstruction of the object wave the developed hologram is again illuminated with the reference wave $\mathbf{r}(x,y) = re^{i2\pi\sigma_r x}$. The hologram $t(x,y)$ acts like a filter and the wave field $\mathbf{u}(x,y)$ directly behind the photographic layer is given by

$$\mathbf{u}(x,y) = \mathbf{r}(x,y) t(x,y). \tag{2.23}$$

With Eqs. (2.15), (2.16), and (2.22a) this becomes

$$\begin{aligned}
\mathbf{u}(x,y) &= (t_0 + \beta\tau r^2)\mathbf{r}(x,y) \\
&\quad + \beta\tau o^2(x,y)\mathbf{r}(x,y) &&: \mathbf{u}_0 \\
&\quad + \beta\tau r^2 \mathbf{o}(x,y) &&: \mathbf{u}_{+1} \\
&\quad + \beta\tau r^2 \mathbf{o}^*(x,y) e^{i4\pi\sigma_r x} &&: \mathbf{u}_{-1}.
\end{aligned} \tag{2.24}$$

This expression describes the effect of a hologram on a light wave during the reconstruction. It is given by four summands which are written in four lines (see also Fig. 2.3b).

The *first* summand refers to the intensity reduction of the reconstruction wave (=reference wave) by the factor $(t_0 + \beta\tau r^2)$ during reconstruction.

The *second* term is small assuming that we choose $o(x,y) < r$ during recording. This term is distinguished from the first term by its spatial variation $o^2(x,y)$. The $o^2(x,y)$ term contains low spatial frequencies which have small diffraction angles and create a so-called halo around the reconstruction wave.

The size of the halo is given by the angular dimension of the object. These first two terms form the zeroth diffraction order in Eq. (2.24).

The *third* expression in Eq. (2.24) denotes the object wave $\mathbf{o}(x,y)$ multiplied with the constant factor $\beta \tau r^2$. An observer who registers this wave in his eye therefore sees the (not present) object. The third term is the most important and represents the first diffraction order. The wave travels divergent from the hologram thus creating a virtual image at the position of the original object. It is a virtual image because the wave is not converging to form a real image. This image cannot be captured on a screen. The intensity (square of the amplitude) of the image does not depend on the sign of β. Therefore it is unimportant whether the hologram is processed "positive" or "negative."

The *fourth* term is essentially the conjugate complex of the object wave $\mathbf{o}^*(x,y)$ and represents the -1st diffraction order. It creates a conjugated real image. The conjugated wave $\mathbf{o}^*(x,y)$ is multiplied with the constant factor $\beta \tau r^2$ as well as the exponential function $e^{i4\pi \sigma_r x}$. The latter means that the wave has roughly twice the angle of incidence (2δ) compared to that of the reference wave (precisely: the sine is twice as large). Since it is the complex conjugated wave the phase changes its sign with respect to $\mathbf{o}(x,y)$. As a consequence the wave $\mathbf{o}^*(x,y)$ travels convergent and forms a real image. As described in Section 2.4, the perspective in a conjugated image is changed in a way that e.g. a concave surface becomes convex. This property is called "pseudoscopic" contrary to the normal image which is called "orthoscopic."

In holography usually two images are created: the normal and the conjugated image [64].

2.4 Conjugated Image

The last section showed that the formation of the conjugate image is quite difficult to understand. But since this image is used in two-stage holography, a more precise description will be presented.

2.4.1 Conjugated Object Wave

To show which properties the conjugate complex object wave $\mathbf{o}^*(x,y)$ has behind the hologram a plane wave with an angle of incidence δ_0 is considered (see Fig. 2.5). In this case the object wave can be written similar to Eqs. (2.16) and (2.17):

$$\mathbf{o}(x,y) = o e^{i2\pi \sigma_0 x} \quad \text{and} \quad \sigma_0 = \sin \delta_0 / \lambda$$

or

$$\mathbf{o}(x,y) = oe^{i\frac{2\pi}{\lambda}x\sin\delta_0}$$ (2.25)

The conjugate complex object wave is formed by changing the sign of the exponent in Eq. (2.25). The negative sign can then be included in the sine function:

$$\mathbf{o}^*(x,y) = oe^{i\frac{2\pi}{\lambda}x\sin(-\delta_0)}.$$ (2.26)

In Fig. 2.5 it becomes apparent that the conjugate complex wave $\mathbf{o}^*(x,y)$ emerges from $\mathbf{o}(x,y)$ by exchanging the angles δ_0 and $-\delta_0$. $\mathbf{o}^*(x,y)$ forms a real image mirrored with respect to the hologram plane.

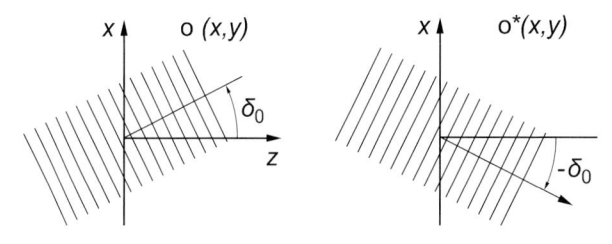

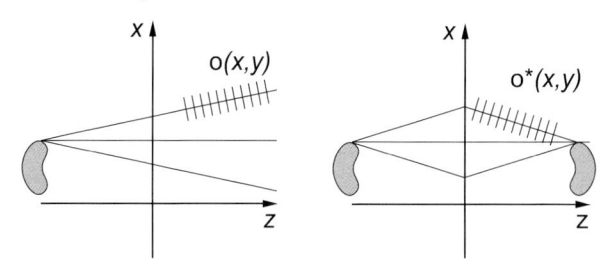

Fig. 2.5 The conjugate complex object wave $\mathbf{o}^*(x,y)$ can be created from $\mathbf{o}(x,y)$ by mirroring at the hologram plane. $\mathbf{o}^*(x,y)$ creates a conjugated pseudoscopic mirror image.

Furthermore, it is noticeable that the three-dimensional images formed by $\mathbf{o}(x,y)$ and $\mathbf{o}^*(x,y)$ have different properties. For an opaque object only the concave inner surface of the image produced by $\mathbf{o}(x,y)$ can be seen, Fig. 2.5. For an image formed by $\mathbf{o}^*(x,y)$ this surface becomes convex. This image with reversed curvatures is called "pseudoscopic." The normal image on the other hand is called "orthoscopic."

2.4.2
Position of the Conjugated Image

The diffraction angle at which the conjugate image forms was already given. It can be calculated from the fourth term $\mathbf{u}_{-1}$ in Eq. (2.24) if for simplification a plane object wave $\mathbf{o}(x,y)$ is assumed. Therefore $\mathbf{o}^*(x,y)$ is also a plane wave. If $\mathbf{o}(x,y)$ has the angle of incidence δ_0 (see Fig. 2.5), Eqs. (2.24) and (2.26) can be written as

$$\mathbf{u}_{-1} = \beta\tau r^2 o e^{i\frac{2\pi}{\lambda}x(\sin(-\delta_0)+2\sin\delta)}. \tag{2.27}$$

Here the relation $e^a e^b = e^{a+b}$ was used. Expression (2.27) represents a wave with the angle of incidence δ_{-1} with

$$\sin\delta_{-1} = \sin(-\delta_0) + 2\sin\delta. \tag{2.28}$$

As already stated the conjugated image with $\delta_0 = 0°$ (object wave impinging perpendicular on the hologram) appears at an angle of

$$\sin\delta_{-1} = 2\sin\delta;$$

δ is the angle of the reference and reconstruction wave.

2.4.3
Reversal of the Reconstruction Wave

In two-stage holography also the real conjugated pseudoscopic image is used, Fig. 2.3b. But sometimes the image position is geometrically unfavorable and it is reconstructed in a different manner, namely be reversing the direction of the reconstruction wave, Fig. 2.3c. The same effect can be achieved by turning the hologram by 180° (around an axis perpendicular to the plane of paper).

The reversal of a plane reconstruction wave like in Eq. (2.16)

$$\mathbf{r}(x,y) = re^{i\frac{2\pi}{\lambda}x\sin\delta}$$

means that δ is replaced by $180° + \delta$. From the relation $\sin(180° + \delta) = -\sin\delta$ follows that the reversed reconstruction wave is described by $\mathbf{r}^*$, Fig. 2.5. By reversing the reconstruction wave the reconstruction described by Eq. (2.23) changes to

$$\mathbf{u}'(x,y) = \mathbf{r}^*(x,y)t(x,y). \tag{2.23a}$$

Instead of Eq. (2.24) the following equation results:

$$\begin{aligned}\mathbf{u}'(x,y) = &(t_0 + \beta\tau r^2)\mathbf{r}^*(x,y) \\ &+ \beta\tau o^2(x,y)\mathbf{r}^*(x,y) &&: \mathbf{u}'_0 \\ &+ \beta\tau r^2 \mathbf{o}(x,y)e^{i4\pi\sigma_r x} &&: \mathbf{u}'_{+1} \\ &+ \beta\tau r^2 \mathbf{o}^*(x,y) &&: \mathbf{u}'_{-1}\end{aligned} \tag{2.24a}$$

The terms $\mathbf{u}'_0$, $\mathbf{u}'_{+1}$, and $\mathbf{u}'_{-1}$ can be interpreted analogous to the last section. The results are shown in Fig. 2.3c.

The *first* and *second* terms form $\mathbf{u}'_0$ in analogy with $\mathbf{u}_0$ in Section 2.3 with the difference that the direction of the reference wave has been reversed.

The *third* term ($\mathbf{u}'_{+1}$) represents the object wave $\mathbf{o}(x,y)$. The direction though is angled by roughly -2δ due to the exponential function. According to Fig. 2.3c a virtual orthoscopic image is formed.

The *fourth* term ($\mathbf{u}'_{-1}$) is of particular practical interest. It describes the conjugated object wave $\mathbf{o}^*(x,y)$. This wave produces an image at the location where the object was placed originally. The image is real but pseudoscopic since it was generated with the conjugated wave. The reversal of the reconstruction wave therefore leads to a real pseudoscopic image which can be used as a new object in two-stage holography.

2.5
Spatial Frequencies

In holography two images appear, the virtual ($\mathbf{o}(x,y)$) and the real ($\mathbf{o}^*(x,y)$) image. In addition to that the reconstruction wave (=reference wave) passing the hologram exists during the reconstruction. The experimental setup needs to be chosen in a way that all three waves are separated in space after a certain distance, Fig. 2.3b. To do so the angle of incidence δ of the reference wave has to be large enough. The diffraction angle under which the three waves diverge is given by the lattice spacing in the hologram. The reciprocal value of this spacing is called the spatial frequency σ; it measures the number of lines per unit length.

If a plane wave, e.g. the reference wave, impinges on the hologram at an angle δ (see Fig. 2.4), a so-called moving grating is formed. Its spatial frequency is

$$\sigma = \sin\delta/\lambda. \tag{2.17}$$

The amplitude of the reference wave in the plane of the hologram is given by

$$\mathbf{r}(x,y) = re^{(i2\pi\sigma_r x)}. \tag{2.16}$$

A plane object wave can be described as follows with the amplitude being constant in this case:

$$\mathbf{o}(x,y) = o(x,y)e^{i2\pi(\pm\sigma_0)x}. \tag{2.29}$$

Here it is assumed that the object wave in the average impinges perpendicular on the hologram. The maximal angle of incidence on the hologram is $\pm\delta_0$, which leads to a spatial frequency spectrum within $\pm\sigma_0$.

Therefore the wave field during reconstruction in Eq. (2.24) can be written as

$$\mathbf{u}(x,y) = (t_0 + \beta\tau r^2)re^{i2\pi\sigma_r x}$$
$$+ \beta\tau o^2(x,y)e^{i2\pi((\pm 2\sigma_0)+\sigma_r)x} \quad : \mathbf{u}_0$$
$$+ \beta\tau r^2 o(x,y)e^{i2\pi(\pm\sigma_0)x} \quad : \mathbf{u}_{+1}$$
$$+ \beta\tau r^2 o(x,y)e^{i2\pi((\pm\sigma_0)+2\sigma_r)x} \quad : \mathbf{u}_{-1}. \qquad (2.30)$$

The second line does not follow directly from Eq. (2.24) since here only a single object wave was permitted (instead of a spatial spectrum within $\pm\sigma_0$). It seems obvious that during the calculation of $o^2(x,y) = o(x,y)o^*(x,y)$ object waves with different spatial frequencies are mixed with each other. In the extreme case the spatial frequency spectrum of $o^2(x,y)$ is limited by $\pm 2\sigma_0$. In Fourier optics the expression is the so-called autocorrelation function of $o(x,y)$ and $o^*(x,y)$ (see the Appendix C).

The spatial frequencies can be read directly in the exponent of Eq. (2.30). By doing so the sign $\pm$ has to be interpreted in a way that all frequencies within the limits $\pm\sigma_0$ occur. Figure 2.6 shows the spatial frequency spectrum. For the object wave $o(x,y)$ a rectangular behavior was assumed. Therefore $\mathbf{u}_{+1}$ and $\mathbf{u}_{-1}$ have the same envelope. The frequency spectrum which constitutes the halo around the reconstruction wave (=reference wave) is twice as wide. For a good holographic recording all three diffraction orders have to be sufficiently separated. The condition that has to be satisfied is

$$\sigma_r \geq 3\sigma_0. \qquad (2.31)$$

Since the spatial frequency is governed by the angle between the normal and the wave direction ($\sigma_r = \sin\delta/\lambda$) the angle of the reference wave has to be large enough during the recording of the hologram:

$$\sin\delta \geq 3\sin\delta_0. \qquad (2.32)$$

δ_0 denotes the maximal allowed angle for the beam impinging on the hologram.

2.6
Diffraction Grating and Fresnel Lens

Holography deals with different aspects of interference optics. Holograms can especially explained by diffraction gratings and Fresnel zone plates [3, 4]. It is therefore worth to understand the principles of holography from different points of view.

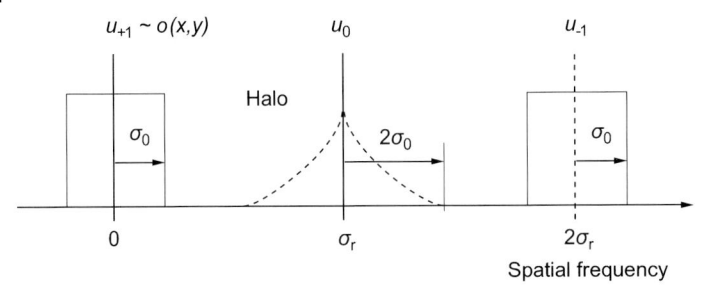

Fig. 2.6 Spatial frequency spectra of the reconstructed object wave u_{+1}, the reference wave during reconstruction u_0 including halo and of the conjugated image u_{-1}. (u_0, $u_{\pm 1}$ correspond to the zeroth and first diffraction orders, respectively.)

2.6.1
Diffraction Grating

In this section the interference, i.e., the superposition, of two plane coherent waves will be calculated. As mentioned in Section 1.2 this procedure can be thought of as holography of a plane object and reference wave. The angles of incidence are denoted as δ_0 and δ (in this case both of them are positive!). Two points P_1 and P_2 are chosen that correspond to the location of two neighboring maxima (or minima) with the spacing d_g (=1/spatial frequency σ). The difference in optical path lengths ($\Delta_0 + \Delta$) according to Fig. 2.7 is

$$\Delta = d_g \sin \delta \quad \text{and} \quad \Delta_0 = d_g \sin \delta_0. \tag{2.33}$$

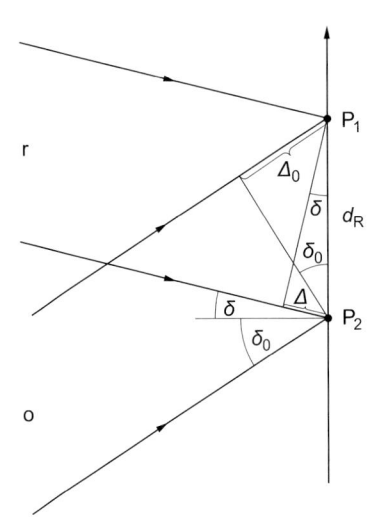

Fig. 2.7 Formation of a holographic diffraction grating.

Using $\Delta_0 + \Delta = \lambda$ the fringe spacing d_g and the spatial frequency σ become

$$d_g = \frac{1}{\sigma} = \frac{\lambda}{\sin\delta + \sin\delta_0}. \tag{2.34}$$

In Section 2.3 it was derived that the intensity transmission of the grating is $\cos^2$- (or $\sin^2$-)like.

The so fabricated hologram, a diffraction grating with the spacing d_g, shall be illuminated with a light wave impinging with an angle α. The maxima of the diffracted light occur at an angle β according to

$$d_g = \frac{N \cdot \lambda}{\sin\alpha + \sin\beta}, \tag{2.35}$$

where N denotes the order of the spectrum. A $\sin^2$- or $\cos^2$-grating only produces the orders $N = 0$ and ± 1. When illuminating it at an angle of $\alpha = \delta$ the first diffraction order is formed at $\beta = \delta_0$. This means that the object wave was reconstructed. The conjugated image is obtained for $N = -1$. The principles of holography are thus contained in the properties of diffraction gratings.

2.6.2
Fresnel Zone Lens

The above-described grating is the hologram of a plane object wave as it is produced in the limit by a far away point. Points of objects close to the hologram reflect or emit spherical waves. Holograms of such object waves have been known for long as "Fresnel zone lenses."

The point P which represents the object is located at the distance z_0 from the photographic layer (Fig. 2.8a). It emits a spherical wave. Additionally a plane reference wave **r** falls onto the layer. First of all it can be noted that the interference pattern consists of concentric circles. For all points that have the same distance from the center of the photographic plate the incoming waves have the same phase. And secondly the path difference between the two interfering waves increases by one wavelength λ from one ring to the other (and the phase difference increases by 2π). The path difference in the center can be taken to be zero. For the kth ring this results in the path difference $k\lambda$, so that the ring radius can be written as

$$r_k^2 = (z_0 + k\lambda)^2 - z_0^2 = 2z_0 k\lambda + k^2\lambda^2. \tag{2.36}$$

When illuminating this set of concentric rings (Fig. 2.9) with parallel coherent light a real and a virtual image point are created (Fig. 2.8b). In terms of holography, Fig. 2.8b represents the reconstruction which is calculated in the following.

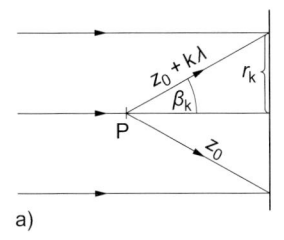

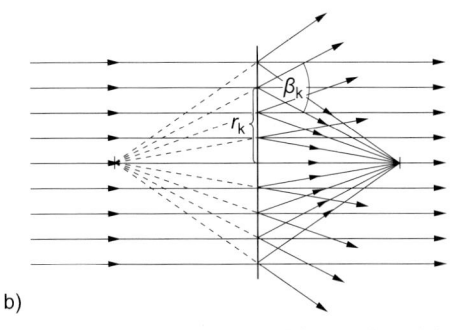

Fig. 2.8 Hologram of a point: (a) recording of the hologram and formation of a Fresnel zone plate and (b) reconstruction and creation of a real and a virtual image point.

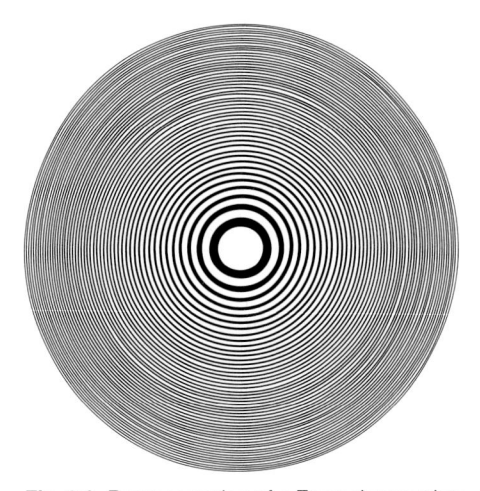

Fig. 2.9 Representation of a Fresnel zone plate. In the ideal case the transition from bright to dark is $\sin^2$-like.

The distance between neighboring rings is calculated from Eq. (2.36) as follows:

$$\Delta r_k = \frac{z_0 \lambda + k\lambda^2}{r_k} \quad \left(= \frac{1}{\sigma} \right). \tag{2.37}$$

Each small area of the zone lens can be interpreted as a regular diffraction grating. The zeroth order diffraction is the weakened illumination beam. Additionally, for a sine-like grating diffraction of the order $N = \pm 1$ occurs at the following angles (see Eq. (2.35) for perpendicular incidence $\alpha = 0$):

$$\sin \beta_k = \pm \frac{\lambda}{\Delta r_k} = \frac{r_k}{z_0 + k\lambda}. \tag{2.38}$$

Equation (2.38) can also directly be calculated from Fig. 2.8b. The deflection angle increases with the distance from the hologram axis. By extensive but elementary trigonometric calculations it can be shown that the beams are intersecting real and virtual at the distance z_0 from the hologram plane as shown in Fig. 2.8b. Therewith it is proved that the hologram of a single point represents a Fresnel zone lens. During reconstruction the first order of diffraction forms a spherical wave which creates an image point at the distance z_0 in front of the hologram. The -1st order of diffraction is a divergent spherical wave with a virtual image point at the distance z_0 behind the hologram.

Problems

Problem 2.1 Calculate the visibility of an interference pattern, the reference wave being four times as intense than the object wave.

Problem 2.2 Verify that using the object wave for reconstruction in Eq. (2.23), the reference wave, $r(x, y)$ is reconstructed.

Problem 2.3 Verify Eq. (2.31) from Eqs. (2.29) and (2.30).

Problem 2.4 What is the meaning of conjugated wave?

Problem 2.5 Verify Eq. (2.37) by differentiating Eq. (2.36).

Problem 2.6 A Fresnel lens has a focal length of 10 cm. Calculate the diameter of the 1st and 100th ring ($\lambda = 600$ nm).

3
Fundamental Imaging Techniques in Holography

The recording of a hologram is performed by bringing object and reference wave to interference. The information of the object wave is completely contained in the resulting interference pattern. During the manufacturing of holograms both waves can be spatially superposed in different ways. Also the photographic layer can be placed in different positions within of the field of interference. From these parameters multiple holographic methods can be deducted each having very specific properties. This chapter only deals with direct methods which do not use any intermediate images generated by lenses or additional holograms before recording. The generation of image-holograms which use an intermediate image of the object is explained in Chapter 4.

The superposition of an object wave of a point O with a reference wave is shown in Fig. 3.1 (see also Section 2.1). In Fig. 3.1a the reference wave emanates as a spherical wave from the point source R; in contrast Fig. 3.1b shows a plane wave as a reference wave, i.e., point R is shifted leftward to infinity. Different holographic methods result depending on the position of the holographic layer within the interference field during the recording:

- In-line hologram after Gabor (Fig. 3.1, position 1)
- Off-axis hologram after Leith and Upatnieks (positions 2, 2', 2'')
- Fourier hologram (position 3)
- Fraunhofer hologram (position 4)
- Reflection hologram after Denisyuk (positions 5, 5')

3.1
In-Line Hologram (Gabor)

The technique of straightforward holography developed by Gabor places the light source and the object on an axis perpendicular to the holographic layer

Holography: A Practical Approach. Gerhard K. Ackermann and Jürgen Eichler
Copyright © 2007 WILEY-VCH Verlag GmbH & Co. KGaA, Weinheim
ISBN: 978-3-527-40663-0

3 Fundamental Imaging Techniques in Holography

a) Spherical reference wave

b) Plane reference wave

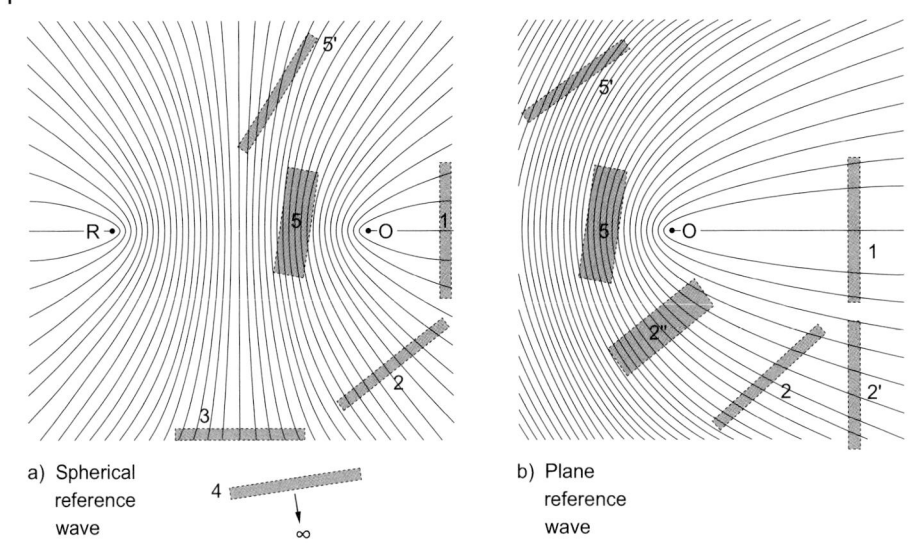

Fig. 3.1 Interference fringes during the recording of holograms of an object point O. 1 In-line hologram after Gabor (thin); 2, 2′ Off-axis hologram after Leith–Upatnieks (thin); 2″ Off-axis hologram (thick); 3 Fourier hologram (lensless); 4 Fraunhofer hologram; 5 Reflection hologram after Denisyuk (thick); 5′ Reflection hologram (transition to thin).

as shown at position 1 in Figs. 3.1a and b. Only transparent objects can be considered. The object wave is represented by the part of the light diffracted by the object whilst the undiffracted part serves as the reference wave (Fig. 3.2a). In principle, both plane and spherical reference waves can be used.

If an axial point O is chosen as an object emitting a spherical wave the resulting hologram for a plane reference wave is a Fresnel zone lens. (Using spherical reference waves results in similar Fresnel lenses.) From Section 2.6 the disadvantage of in-line or straightforward holograms is obvious: during reconstruction the hologram is illuminated with a plane reference wave according to Fig. 3.2b. Since it represents a zone lens a virtual image point is formed at the original point and additionally a real image point appears at the same distance to the right of the hologram.

The phenomenon also holds for extended objects which can be divided into single points. During observation the two images lying on the same axis interfere which leads to image disturbances (Fig. 3.2b). Moreover, the observer looks directly into the reconstruction wave. Because of these disadvantages this form of holography is only of historical interest. A single laser beam is used for the recording which constitutes the object and reference wave without splitting the beam. Such techniques are called "single beam" holography. The hologram is illuminated from the backside when observing the image;

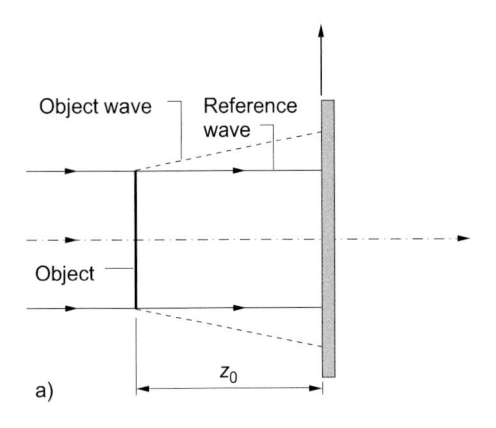

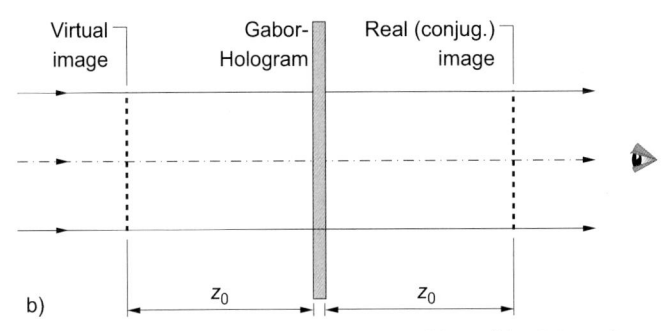

Fig. 3.2 In-line holography for transparent objects (Fig. 3.1, position 1): (a) recording of the hologram and (b) reconstruction of the hologram.

it is called a "transmission hologram." From Fig. 3.1 it is apparent that the interference lines are perpendicular to the light sensitive layer and have a relatively large distance. For common film layers the hologram can be classified as "thin" particularly since the grating spacing is relatively large. The difference between the so-called "thin" and "thick" holograms will be explained in Chapter 6.

3.2
Off-Axis Hologram (Leith–Upatnieks)

It turns out to be more favorable to shift either the holographic layer or the object sideways and use position 2 or 2′ according to Fig. 3.1. This technique was developed by Leith and Upatnieks and has prevailed for many applications. Laser beam, object, and hologram are not on the same axis anymore.

The hologram represents the outer area of a fresnel zone lens (at least in setup 2′). Again a virtual and a real image are formed during reconstruction. For position 2′ of the hologram both images lie at the same position as with Gabor holography (position 1). The advantage of off-axis holography is that both images are positioned at different angles regarding hologram 2′. Therefore the images do not interfere during observation and image disturbances are avoided.

A detailed description of off-axis holography was given in Sections 2.3 and 2.4 and Fig. 2.3. The setup for off-axis holography is repeated in Fig. 3.3. By tilting the reference wave (or shifting the object) it is achieved that the three diffraction orders, namely the image, the conjugated image, and the illumination wave, are spatially separated. This has the advantage that also holograms of opaque objects can be produced since the reference wave is not obstructed by the object.

The setups according to Figs. 3.2 and 3.3 generate transmission holograms. The formation of a "thin" or "thick" grating depends on the thickness of the holographic layer, the grating spacing and the direction of the grating planes. In practice, mostly thin transmission holograms occur. They can only be reconstructed with monochromatic light since every color generates a different diffraction angle for the image position. The use of white light generates a totally blurred image due to these chromatic aberrations.

In thick transmission holograms Bragg reflection only occurs for the wavelength used during recording. This makes it possible to use white light for reconstruction. The effect gets stronger with increasing tilt angle of the grating planes. With 45° tilt the reconstruction can be done with almost white light. Although this means grazing incidence for the reconstruction wave, for the conjugated image the Bragg condition is not satisfied and it does not appear. The diffraction efficiency of the virtual image is therefore increased (see Appendix B). If the photographic layer is placed at position 2″ (Fig. 3.1), a thick hologram is formed more likely than in positions 2 and 2′ since the grating lines are closer together.

In Fig. 3.1 it is not obvious whether off-axis holograms are produced using a single beam or a multiple beam technique. In principle, both are possible. In split-beam holography the reference wave is separated from the illumination of the object by a beam splitter.

3.3
Fourier Hologram (Lensless)

If the object O and the point light source R are within the same plane parallel to the hologram so called "Fourier holograms" are generated (Fig. 3.1, posi-

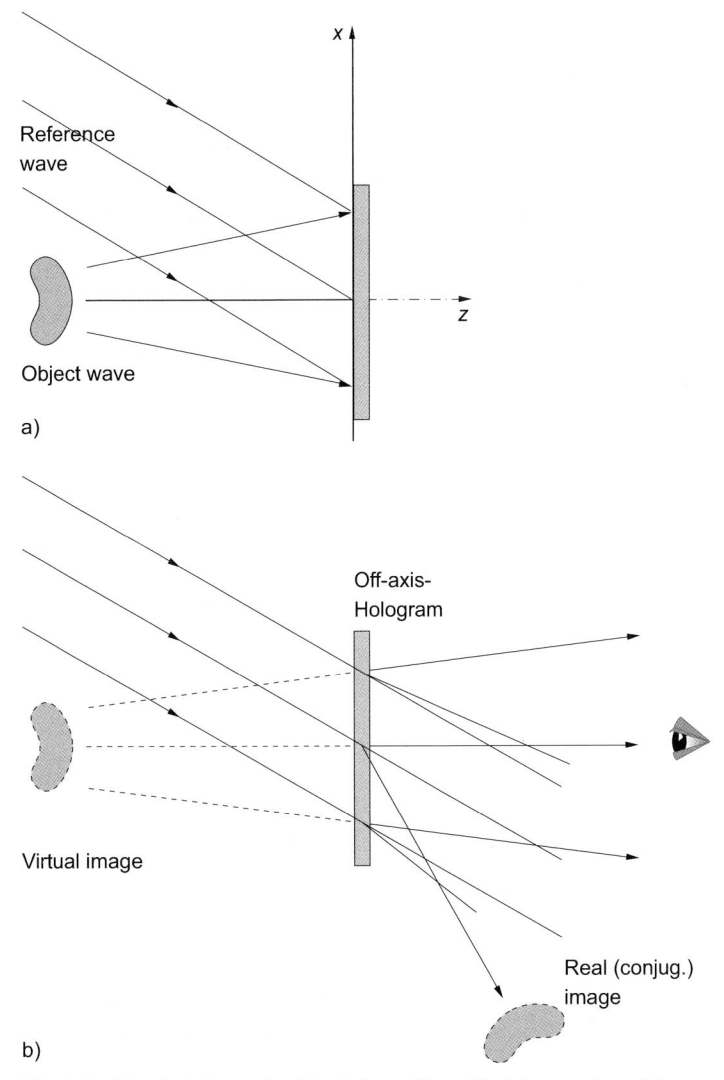

Fig. 3.3 Off-axis holography (Fig. 3.1, positions 2): (a) recording of the hologram and (b) reconstruction of the hologram.

tion 3). This geometric condition can only be satisfied for plane objects. In a Fourier hologram the interference fringes appear as a set of hyperbolas whilst especially in in-line holograms circular sets in the form of Fresnel zone lenses appear.

A schematic setup for the recording and reconstruction of lensless Fourier holograms is shown in Fig. 3.4. Like in all thin holograms two (real) images appear during reconstruction. The regular image is at the position of the original object; the conjugated one appears in the same plane parallel to the hologram. The point light source R is the center of point symmetry for the two images (Eq. (2.28)). The special properties of these holograms are treated in Section 4.5 in connection with Fourier holograms using lenses.

3.4
Fraunhofer Hologram

Fourier holograms are formed by the superposition of spherical waves whose centers have the same distance from the holographic layer. If the layer is moved far away as in Fig. 3.1 the centers depart and in the limit plane waves are created (position 4); one talks of "Fraunhofer holograms."

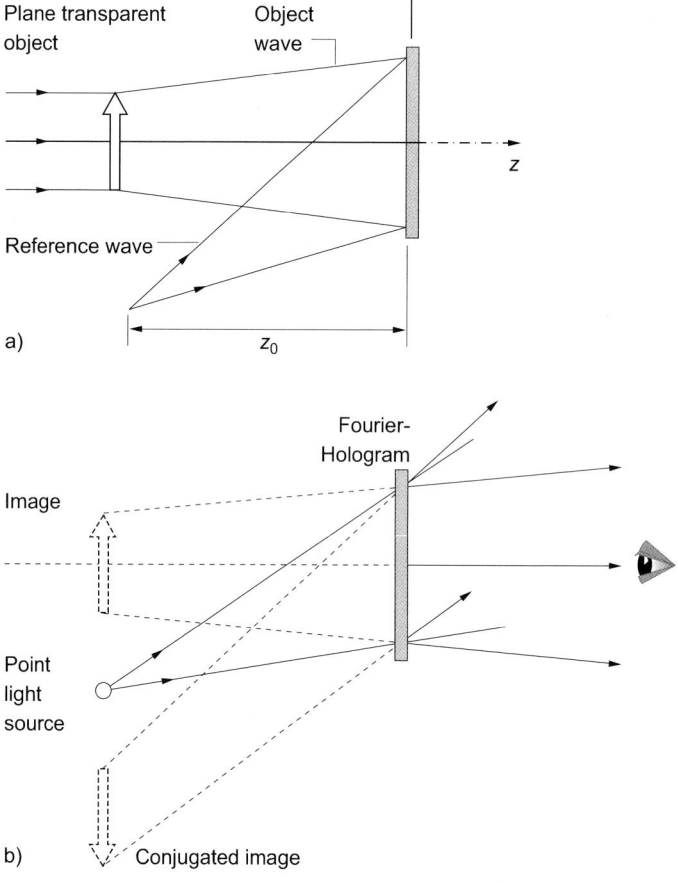

Fig. 3.4 Lensless Fourier hologram (Fig. 3.1, position 3): (a) recording of the hologram and (b) reconstruction of the hologram.

Often these holograms are defined in a different way. Diffraction in the far field is known as "FFraunhofer diffraction" for a long time. (For very small objects, i.e., aerosols, the far field is already given at a distance of a few millimeter). Holograms of small objects in the far field using a plane reference wave are therefore also called "Fraunhofer holograms."

This hologram type is especially used for the measurement and investigation of aerosols. A setup for this is shown in Fig. 3.5. The object with radius r_0 has to be so small that a diffraction pattern will appear in the far field. The condition for the distance object/hologram is $z_0 \gg r_0^2/\lambda$.

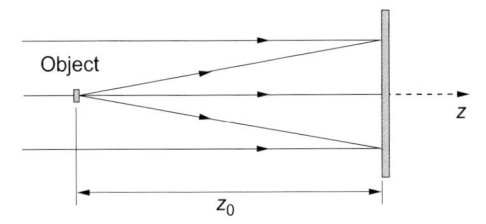

Fig. 3.5 Recording of a Fraunhofer hologram (Fig. 3.1, position 4).

Figure 3.5 represents the Gabor holography with the condition of diffraction being present in the far field. The light of the primary image is spread over such a large area in the conjugated image that it appears as a weak even background.

3.5 Reflection Hologram (Denisyuk)

Until now thin holograms were presented in this chapter at which the object and reference wave impinge from the same side on the photographic layer. Holograms whose images are reconstructed in reflection are of large importance especially in the field of graphics and art. In this case the reference wave – and later the reconstruction wave – has to impinge from the observer's side onto the hologram. The object wave in this type of recording impinges on the hologram from the opposite side. This is equivalent to positions 5 and 5' in Fig. 3.1.

Of importance is the setup after Denisyuk in which the holographic layer is positioned across between the light source and the object. This results in the interference planes being almost parallel to the light sensitive layer. The distance of the grating planes when using a He–Ne or ruby laser is $\lambda/2 \approx$ 0.3 µm. Therefore, for a typical layer thicknesses of around 6 µm, almost 20 grating planes fit into the light sensitive layer. So this system behaves like a thick grating.

The common diffraction theory (Sections 1.2 and 2.6 has to be modified for reflection holograms. Thick gratings exhibit a totally different behavior (Chapter 6). During reconstruction the illumination wave which is ideally identical to the reference wave is reflected at the grating planes. The virtual image of the object appears in the reflected light. Interference effects appear during the mirroring which lead to Bragg reflection. If white light is used for illumination only the wavelength used for the recording is reflected due to the Bragg effect. Therefore a sharp monochromatic image appears although white light is used for reconstruction. This is the advantage of thick reflection holograms which are called "white light holograms."

The techniques for recording and reconstruction of reflection holograms are shown in Fig. 3.6. An extension of these techniques to two-step methods is described in Chapters 4 and 10.

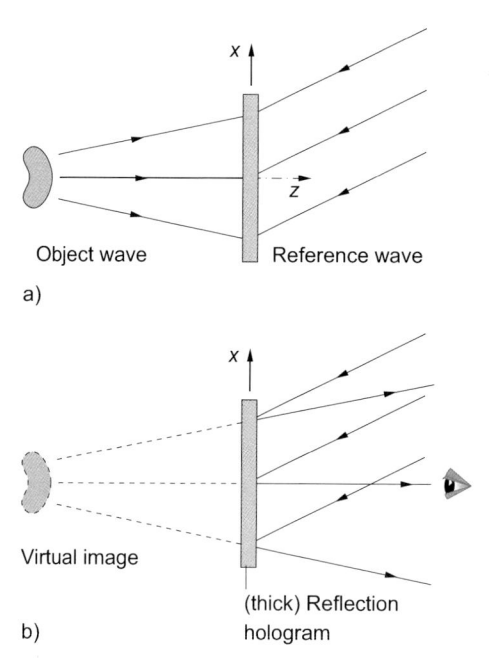

Fig. 3.6 Reflection hologram (Fig. 3.1, positions 5): (a) recording of the hologram and (b) reconstruction of the hologram.

If the holographic layer is shifted to position 5′ during recording (see Fig. 3.1) the grating constant is increased and the grating planes lie inclined within the layer. In this case the Bragg effect is less distinct and the grating cannot be regarded as "thick" anymore. To distinguish this technique from off-axis holography it is often called "holography with an inverted reference beam."

Problems

Problem 3.1 What are the main disadvantages of a Dennis Gabor in-line-hologram?

Problem 3.2 What is the difference in the positions of reference and object waves with respect to the holographic plate for a transmission and a reflection hologram during exposure and reconstruction?

Problem 3.3 Give the position of the interference fringes within the emulsion for transmission- and reflection-holograms. (Fig. 3.1, No. 2 and No. 5, respectively)

Problem 3.4 What is the advantage of off-axis-holography?

Problem 3.5 Is a "Single-Beam" exposure in contradiction to the principle of holography that two waves have to interfere in order to store an image?

4
Holograms of Holographic Images

In the preceding chapter it was assumed that the object wave falls directly onto the hologram without any further imaging during recording. One characteristic of this setup is that the normal (orthoscopic) image appears virtual and behind the hologram during observation. The topic of this chapter are holograms where the object wave is imaged using a lens before recording or emerges from a "master hologram." An image created by a lens or a hologram serves as an object; therefore these holograms are called "image holograms." By two-step manufacturing methods real (orthoscopic) images can be created that can be observed in front of the hologram. This technique is especially interesting for graphical and artistic applications as well as holographic displays.

4.1
Image-Plane Hologram

It has a lot of advantages to record the real image of an object instead of the object itself. For image-plane holograms the object is imaged into the plane of a hologram by a large lens (Fig. 4.1). During the recording and also the reconstruction the real image of extended objects is partly in front of and behind the hologram.

Due to the hologram plane being in the middle of the image the differences in path lengths are smaller than those in other techniques. Hence minimal demands are made regarding the coherence of the light source. If the depth of the object is small even "white" light sources can be used. Another advantage is that image-plane holograms are relatively bright and brilliant, though the observation angle is limited by the lens aperture.

4.2
Transmission and Reflection Hologram in Two Steps

Real images can also be created without using any lenses with the help of holographic two-step methods.

Holography: A Practical Approach. Gerhard K. Ackermann and Jürgen Eichler
Copyright © 2007 WILEY-VCH Verlag GmbH & Co. KGaA, Weinheim
ISBN: 978-3-527-40663-0

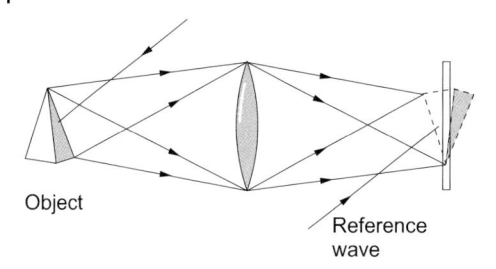

Fig. 4.1 Recording of an image-plane hologram.

In the first step an off-axis hologram is created according to Fig. 4.2a which is called "master" or "H1 hologram." The reconstruction creates a virtualorthoscopic image and a real pseudoscopic image. The real image delivers the object wave for the second step when the so-called H2 hologram is recorded. Thus a hologram of a holographic image is created in two-step holography. Using this technique it is possible to create real orthoscopic, i.e., normal, images since the pseudoscopic image of a pseudoscopic image is orthoscopic. Figure 4.2 shows the manufacturing of a two-step transmission hologram. For the creation of a white light reflection hologram the direction of the reference wave during the recording of the H2 hologram has to be inverted; the same holds true for the reconstruction wave. An disadvantage of the setup shown in Fig. 4.3 is the laterally shifted position of the orthoscopic image which hampers the observation.

A frequently used alternative to the creation of real holographic images is shown in Fig. 4.3. In the first step a regular off-axis transmission hologram is created, similar to Fig. 4.2a. In the example shown the reference wave impinges at an angle and the object stands in front of the light sensitive layer. In the second step the master hologram is rotated by 180° (around an axis perpendicular to the plane of paper). For spatial orientation a marker is added to the hologram in Fig. 4.3. This rotation has the same effect as the inversion of the direction of the reconstruction wave (see Section 2.4).

The reconstruction in the second step generates a real pseudoscopic image which is upside down. The light sensitive layer of the H2 hologram is often positioned in a plane that runs right through the image. Thus the real pseudoscopic image of the master hologram is recorded in the H2 hologram. Since the propagation directions of object and reference wave point from different directions onto the hologram a white light reflection hologram is generated. For the reconstruction of the recorded information the H2 hologram is again rotated by 180°. Hence a real pseudoscopic image appears (according to Section 2.4). Though relating to the object the real image is orthoscopic.

Two-step reflection holograms are mostly found in holographic art galleries. The illumination can be done using white light, e.g., with 12 V halogen lamps.

4.2 Transmission and Reflection Hologram in Two Steps

a) Master hologram

Reference wave
Master
Object
Object wave

b) H2-Hologram

Reconstruction wave
Virtual orthoscopic image
Real pseudoscopic image
Reference wave
H2

c) Reconstruction

Virtual image
Real orthoscopic image
Reconstruction wave
H2

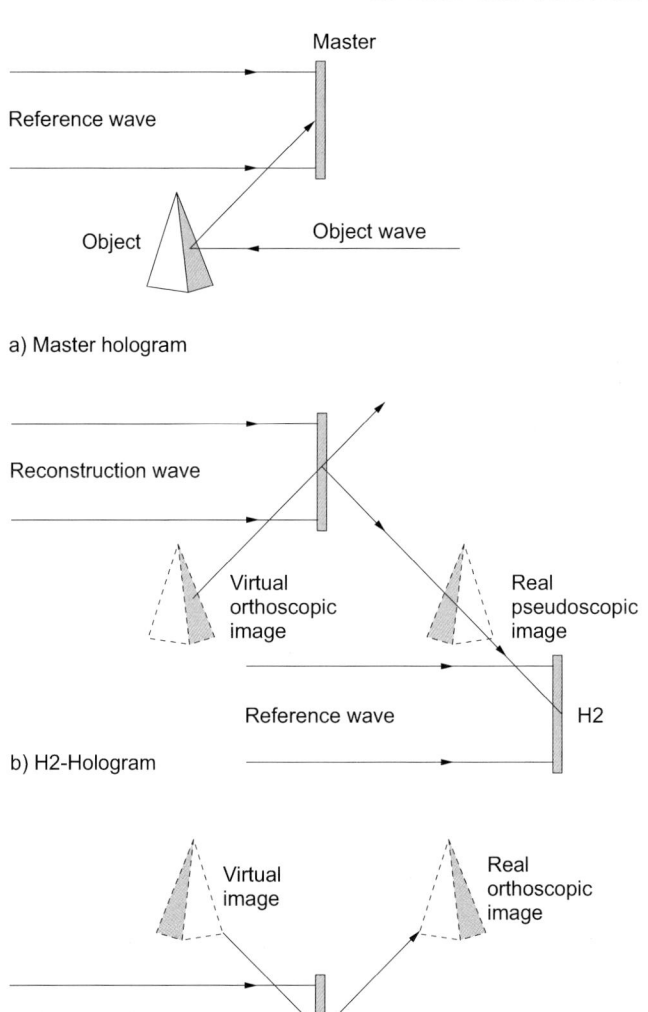

Fig. 4.2 Generation of a real orthoscopic image using two-step holography: (a) recording of a master hologram, (b) creation of a H2 hologram in transmission, and (c) reconstruction of the real image.

The angle of incidence for the light is defined by the direction of the reference wave used during recording of the H2 hologram. Technical details for the manufacturing of this type of holograms are described in Chapter 11.

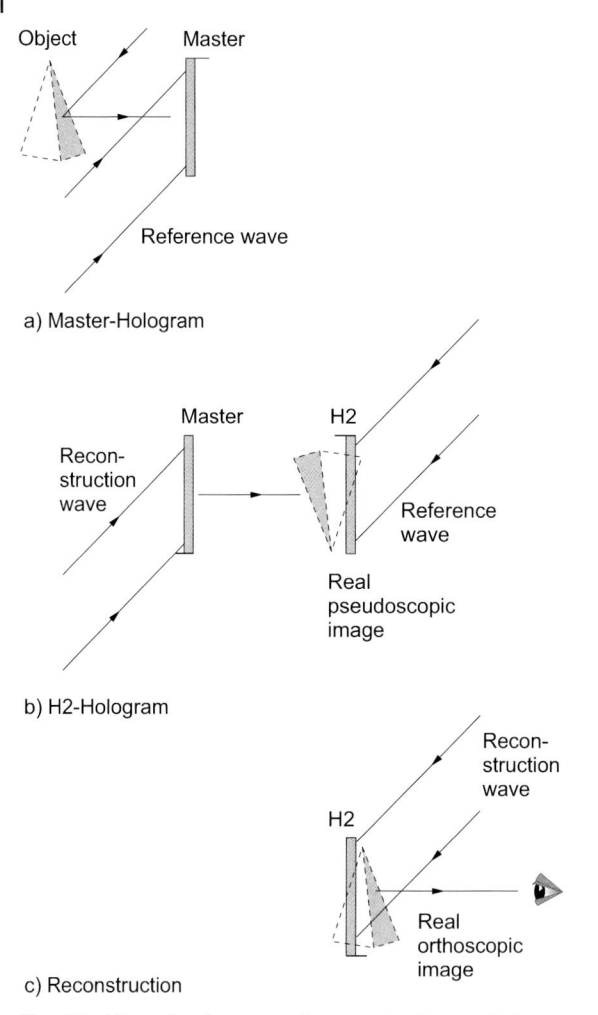

Fig. 4.3 Alternative for generating a real orthoscopic image using two-step holography: (a) recording of a master hologram, (b) creation of an H2 reflection hologram, and (c) reconstruction of the real image.

4.3
Rainbow Hologram

Rainbow holograms can be reconstructed in transmission using white light. Depending on the viewing direction the reconstructed image appears in different colors, exhibiting the whole light spectrum. The technique for the recording of rainbow holograms consists of two steps. In the first step an off-axis hologram is created in the usual manner (Fig. 4.4a). For the reconstruction the hologram is rotated by $180°$ (or the direction of the reference beam is inverted) and thus a real pseudoscopic image is created (Fig. 4.4b).

a) Master-Hologram

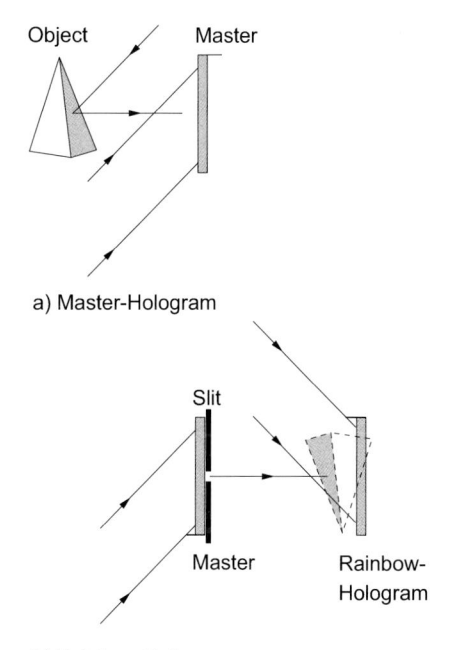

b) Rainbow-Hologram

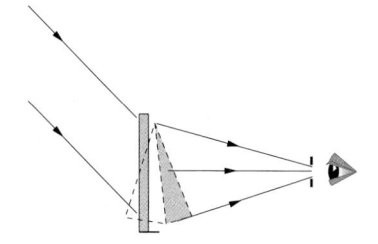

c) Reconstruction with monochromatic light

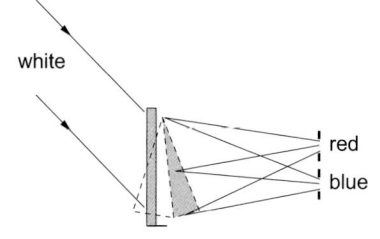

d) Reconstruction with white light

Fig. 4.4 Technique for creating rainbow holograms: (a) recording the master hologram, (b) generating the rainbow hologram using a horizontal slit aperture, (c) reconstruction of the rainbow hologram using monochromatic light, and (d) reconstruction of the rainbow hologram using white light.

Contrary to Fig. 4.3a, a horizontal slit aperture is mounted on the hologram. By doing so information is lost and the reconstructed image lacks a vertical parallax, i.e., the three-dimensional impression in the vertical direction is lost. Usually this will not be noticed by the observer: the reason for that is the eyes being oriented horizontally.

A photosensitive layer is positioned inside the real image and the H2 hologram is created according to Fig. 4.4b. By this process the information of both the pseudoscopic image and of the slit aperture is recorded.

To reconstruct the images of rainbow holograms they are rotated by 180° to create an orthoscopic image from the pseudoscopic one. In Fig. 4.4c the reconstructed image is shown using monochromatic light. The observer looks through a horizontal slit which is the image of the aperture that was used. A high intensity is achieved since the diffracted light is concentrated on this slit. The viewing angle although is limited and the three-dimensional impression exists only in the horizontal direction.

When using white light for reconstruction the image of the horizontal slit appears under a different diffraction angle, see Fig. 4.4d. For each spectral color there exists a different viewing slit. If the observer moves the head in the vertical direction he or she will see the image successively in red, orange, yellow, green and blue, i.e., in the spectral colors or the colors of the rainbow. This explains where the term "rainbow hologram" originates from.

4.4
Double-Sided Hologram

Usually only the information of the front side of a three-dimensional object can be recorded on a plane hologram. With double-sided holograms the hologram can be viewed from two sides (Fig. 4.5) and the reconstructed image shows front and backside of the object.

The production of a double-sided hologram starts with the recording of a transmission master hologram H1 of one side of the object (1). After that a second hologram H2 of the wavefront from the other side of the object (2) is recorded. This one is not developed at first and a latent reflection hologram is created.

Starting from this hologram H2 the third step consists in creating a double-sided hologram. In doing so a pseudoscopic real image of side (1) of the object is generated from the master hologram by inverting the direction of the reference wave. On the second hologram H2 a second exposure is made and the wavefront from the master is recorded. The direction of the reference wave is different from the first exposure, see Fig. 4.5. For the reconstruction the illumination wave has to be inverted again since the image of side (1) was

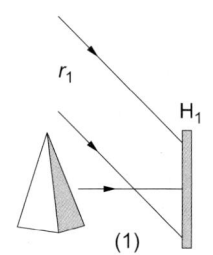

a) Master-Hologram

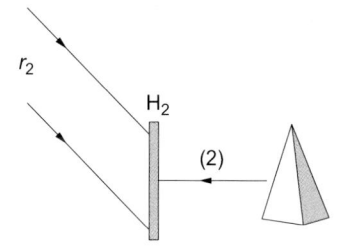

b) Exposure of reverse side (2)

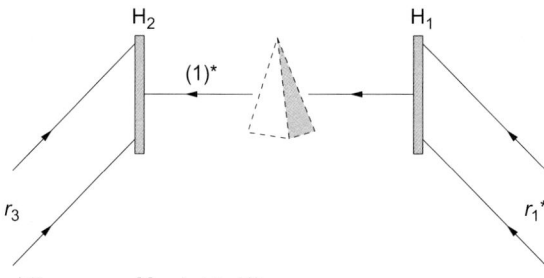

c) Exposure of front side (1)

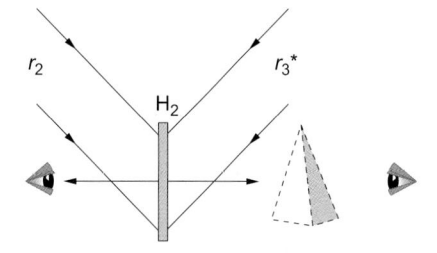

d) Reconstruction of both sides of object

Fig. 4.5 Techniques for creating double-sided holograms: (a) creating a master hologram in transmission of side (1) of the object, (b) creating a second hologram in reflection of side (2), (c) additional recording of the information of side (1) on the second hologram, and (d) reconstruction of the double-sided hologram.

pseudoscopic. The whole procedure is shown in Fig. 4.5. Two independent reflection holograms are obtained which display both sides of the object by a virtual (illumination by r_2) and a real image (illumination by r_3^*).

4.5
Fourier Hologram

Lensless Fourier holograms were already described in Section 3.3. In the area of pattern recognition they are often used with lenses (Section 18.1).

4.5.1
Principle

For producing a Fourier hologram a plane object is placed in the first focal plane of the lens. The reference wave emerges from a point light source in the same plane (Fig. 4.6). The photographic layer is placed in the second focal plane of the lens during recording. The reconstruction is done by illuminating the hologram with an axially parallel plane wave. The hologram is again placed in the first focal plane of a similar second lens. The primary and the conjugated images appear in the second focal plane symmetric to the optical axis, see Fig. 4.6b. The undiffracted reference wave forms an axial "light spot" representing the zeroth diffraction order. It can be shown that the reconstructed image remains stationary when the hologram is shifted in its plane.

4.5.2
Calculation

The calculation of the Fourier hologram can be done according to Fig. 4.6. Assuming that the object is a flat and transparent piece (slide) in the front focal plane of a lens with the object point coordinates x and y. The object is illuminated by a plane wave of coherent laser light. The object wave $\mathbf{o}(x,y)$ (e.g. complex amplitude) is transformed to the Fourier transform $F\{\mathbf{o}(x,y)\}$ (see Appendix C) at the back focal plane of the lens.

$$\mathbf{O}(\xi,\eta) = F\{\mathbf{o}(x,y)\}. \tag{4.1}$$

Here ξ and η are the coordinates (spatial frequencies) in the back focal plane of the lens. The reference wave is emitted by a point source in the front focal plane at $x = -b$ (see Fig. 4.6) and is described by a delta function:

$$\mathbf{r}(x,y) = \delta(x+b,y). \tag{4.2}$$

The Fourier transform of the delta function is

$$\mathbf{R}(\xi,\eta) = e^{-i2\pi\xi b}.$$

4.5 Fourier Hologram

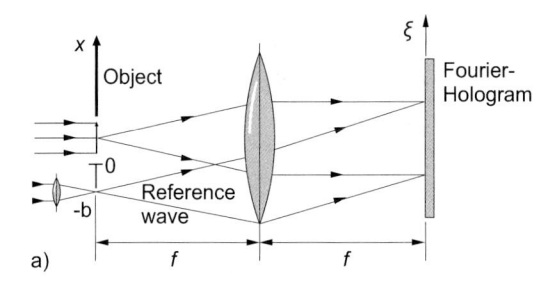

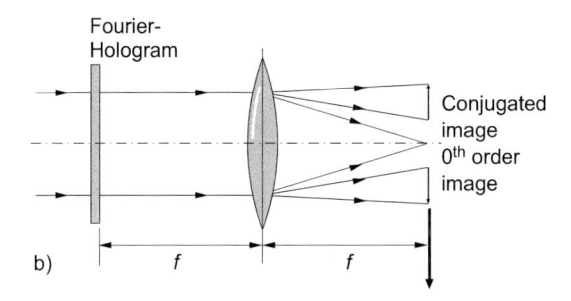

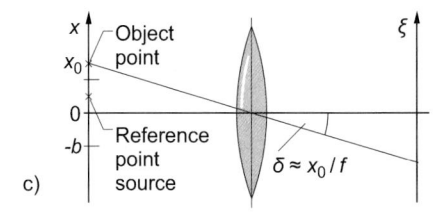

Fig. 4.6 Fourier hologram. (a) exposure, (b) reconstruction, (c) calculating a Fourier hologram (see text).

That means the square of $R(\xi,\eta)$, the intensity, is constant and just 1. Both waves interfere. The resulting intensity at the plane of the holographic plate is given by

$$I = \{O(\xi,\eta) + R(\xi,\eta)\}^2$$
$$I = 1 + |O(\xi,\eta)^2| + O(\xi,\eta) \cdot R^*(\xi,\eta) + O^*(\xi,\eta) \cdot R(\xi,\eta). \tag{4.3}$$

The resulting amplitude transparency of the holographic plate again is a linear function of intensity as assumed already in Eq. (2.21). If then the processed holographic plate is placed in the front focal plane of the lens and illuminated with a plane reference wave of unit amplitude, it follows for the wave, reconstructed by the plane wave in accordance with Eqs. (2.23) and (2.24) for a constant reconstruction wave of amplitude 1:

$$U(\xi,\eta) = t_0 + \beta T I(\xi,\eta). \tag{4.4}$$

The Fourier transform of Eq. (4.4) in the back focal plane of the lens is

$$\mathbf{u}(x,y) = \mathbf{F}\{\mathbf{U}(\xi,\eta)\}$$
$$\mathbf{u}(x,y) = (t_0 + \beta\tau) \cdot \delta(x,y)$$
$$\quad + \beta\tau \cdot \mathbf{o}(x,y)(\bullet)\mathbf{o}(x,y)$$
$$\quad + \beta\tau \cdot \mathbf{o}(x-b,y)$$
$$\quad + \beta\tau \cdot \mathbf{o}^*(-x+b,-y). \tag{4.5}$$

The reconstruction of the wave results in four terms similar to Eq. (2.24): in a point source (zeroth order) surrounded by a halo (second summand of Eq. (4.5), the autocorrelation function, see Appendix C) and two symmetrically positioned real images. This is the speciality of Fourier transform holograms: there are two real images reconstructed. The reconstructed images will not move if the hologram is moved within the (ξ,η)-plane. This is due to the fact that the Fourier transform of such a movement will result in a phase factor, only, which will not affect the intensity distribution of the reconstructed images. The conjugate pseudoscopic image can be discriminated from the orthoscopic image because it is upside down.

Problems

Problem 4.1 What is the difference between an orthoscopic and a pseudoscopic image? In this respect, what is the difference between the virtual and the real image?

Problem 4.2 An image plane hologram can be reconstructed using white light. What is the reason?

Problem 4.3 During exposure of a rainbow hologram does diffraction of the slit plays an important role?

Problem 4.4 Image plane holograms are called full aperture holograms in contrast to slit aperture holograms (rainbow holograms). Comment the difference and the similarity of both types.

Problem 4.5 Show, that the Fourier transform of the delta function $r(x,y) = \delta(x+b,y)$ is $R(\xi,\eta) = e^{-i2\pi\xi b}$.

5
Optical Properties of Holographic Images

The original object wave is recreated if the light source during reconstruction has the same spatial position and wavelength as the one used during the recording. But an errorless recording of the interference pattern with no aberrations in the photographic layer is required. In practice, this ideal case is not achievable. This chapter describes the properties of the reconstructed image against different parameters of the optical system.

5.1
Hologram of an Object Point

First of all the equations for the geometrical calculation of the holographical imaging of an object point are derived. From that the position of the reconstructed image and the scale of the image can be calculated for all cases.

5.1.1
Image Equations

The image equations are derived using the notation given in Fig. 5.1. The hologram lies in the x,y-plane at $z = 0$. The x,y,z coordinates have the following subscripts: object point (o), reconstructed image (i), reference point source (r), and point light source for reconstruction (c). The grating constants of the hologram can change during processing, i.e., due to shrinking of the gelatin. The variable m denotes the ratio of the distances before (d_g) and afterward (d'_g): $m = d_g/d'_g$. The wavelength during reconstruction, λ', can be different from that during recording, λ, described by the factor $\mu = \lambda'/\lambda$.

For the image position the following equations result [3]:

$$z_i = \frac{m^2 z_c z_o z_r}{m^2 z_o z_r \pm \mu z_c z_r \mp \mu z_c z_o}$$

$$x_i = \frac{m^2 x_c z_o z_r + \mu m x_o z_c z_r \mp \mu m x_r z_c z_o}{m^2 z_o z_r \pm \mu z_c z_r \mp \mu z_c z_o} \tag{5.1}$$

$$y_i = \cdots \text{(exchanging x by y in the last equation)}$$

Holography: A Practical Approach. Gerhard K. Ackermann and Jürgen Eichler
Copyright © 2007 WILEY-VCH Verlag GmbH & Co. KGaA, Weinheim
ISBN: 978-3-527-40663-0

5 Optical Properties of Holographic Images

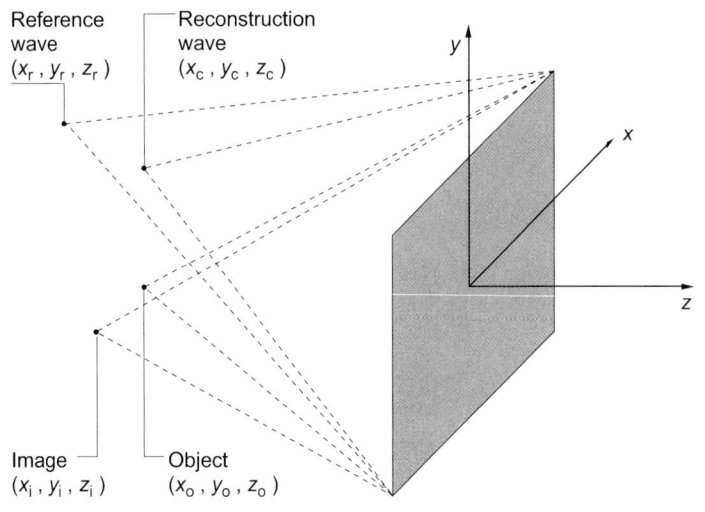

Fig. 5.1 Imaging and reconstruction of an object point according to Eq. (5.1). Subscripts: o = object, r = reference point light source, c = point light source for reconstruction, i = image.

The upper sign applies to the normal image whilst the lower sign applies to the conjugated image. The relation for the image position z_i can be rewritten in a way that it formally resembles the lens equation. Considering only the distances z in the hologram and calculating the reciprocal the result for z_i is

$$\frac{1}{z_i} = \frac{1}{z_c} \pm \frac{\mu}{m^2 z_0} \mp \frac{\mu}{m^2 z_r}. \tag{5.2}$$

According to this relation the hologram can be thought of as an optical element having two focal lengths where one belongs to the normal image and the other belongs to the conjugated image. The dependence of the distances z of the normal and the conjugated image follows from Eq. (5.2):

$$\frac{1}{f} = \frac{1}{z_{in}} + \frac{1}{z_{iu}} \quad \text{and} \quad f = \frac{z_c}{2}. \tag{5.3}$$

This equation also resembles the lens equation; z_{in} denotes the position of the normal image and z_{iu} that of the conjugated image. This equation is independent of the object position and the wavelength used.

5.1.2
Magnification

The lateral or transverse magnification defines the ratio of the image size (dx_i) and the object size dx_o:

$$V_{\text{lat}} = \frac{dx_i}{dx_o} = \frac{dy_i}{dy_o}. \tag{5.4a}$$

Using Eq. (5.1) this results in

$$V_{lat} = \frac{m}{1 \pm \frac{m^2 z_o}{\mu z_c} - \frac{z_o}{z_r}}, \qquad (5.4b)$$

where again the upper sign belongs to the normal image and the lower sign to the pseudoscopic image.

The magnification for the normal image is equal to 1 if the condition for recording and reconstruction are kept the same ($m = 1, \mu = 1, z_r = z_c$). For the real pseudoscopic image $V_{lat} = 1$ results when $m = 1, \mu = 1$, and $z_r = -z_c$. $V_{lat} = 1$ means that either plane waves have to be used or when using converging light that diverging light is needed for reconstruction.

In two-step production of holograms it is often favorable to use small objects. If the master hologram is created with diverging light this results in a magnification in the real pseudoscopic image when the same wave is used for reconstruction $z_r = z_c$. This can be easily seen in Eq. (5.4b) when assuming $m = \mu = 1$ and looking at the lower sign. The magnified image can be recorded in the second hologram. By changing z_c, i.e., the divergence of the wave for reconstruction, the magnification can be varied.

The use of small objects with later magnification has some advantages. The two-step process can create images whose extent can be larger than the coherence length. Furthermore, small objects and the use of large holographic films make it possible to record a large spatial angle and therefore result in a large viewing angle.

5.1.3
Angular Magnification

Objects and images are viewed at a certain angle. If the observer's eye is in the hologram plane the resulting angular magnification is approximately

$$V_\alpha = \frac{d(x_i/z_i)}{d(x_o/z_o)} \qquad (5.5a)$$

and therefore for both images

$$V_\alpha = \frac{\mu}{m}. \qquad (5.5b)$$

5.1.4
Longitudinal Magnification

The magnification in the longitudinal direction is given by

$$V_{long} = \frac{dz_i}{dz_o}. \qquad (5.6a)$$

This results in

$$V_{long} = \pm \frac{V_{lat}^2}{\mu}. \quad (5.6b)$$

Normal and conjugated images have different signs in the longitudinal magnification; the images are therefore orthoscopic and pseudoscopic.

The longitudinal magnification rises quadratically with the lateral one. This causes an aberration in the images which intensifies the spatial impression. Two-step holograms can therefore be used to create images several meters in front of the recording layer.

5.1.5
Image Aberrations

Numerous image aberrations occur if the conditions during reconstruction are not the same as during recording. Additional information on this topic can be found in the literature listed in the references [3, 27, 44].

5.2
Properties of the Light Source

The position of the image point x_i is given by Eq. (5.1) if a monochromatic point light source is used for reconstruction. The shift of x_i by changing the position of the light source x_c can be obtained by derivation:

$$\frac{dx_i}{dx_c} = \frac{z_0 z_r}{z_0 z_r + z_c z_r + z_c z_0}. \quad (5.7)$$

From this equation the blurring of the image by enlarging the light source can be calculated. If the position of the reference and illumination source is the same ($z_r = z_c$) the resulting image blur dx_i can be derived from Eq. (5.7):

$$dx_i = \frac{z_0}{z_c} dx_c. \quad (5.8)$$

The tolerable blur dx_i is governed by the resolution of the eye of about 0.5 mrad. For a distance of 1 m this results in $dx_i \approx 0.5$ mm. Assuming that objects and images have a distance of 5 cm from the hologram ($z_0 = 5$ cm) and the light source is 1 m away ($z_c = 100$ cm), the allowable diameter of the light source is up to 1 cm. Therefore, e.g., a high-pressure Hg lamp can be used as a light source.

5.2.1
Spectral Bandwidth

The influence of the spectral bandwidth of the light source on the reconstructed image can be calculated by assuming plane reference and reconstruction waves ($z_c = z_r = \infty$). It has to be kept in mind that the expressions $x_c/z_c = x_r/z_r$ are finite and describe the direction of the wave. Eq. (5.1) can be simplified to

$$x_i = x_o + \frac{x_c}{z_c} \cdot \frac{z_0}{\mu} - \frac{x_r}{z_r} \cdot z_0$$

$$z_i = \frac{z_0}{\mu} \tag{5.9}$$

with

$$\mu = \frac{\lambda_c}{\lambda_r}.$$

The ratio of the wavelengths of reconstruction and reference wave is defined by μ.

The shift of the image by changing the wavelength of the reconstruction wave, $d\lambda_c$, is given by

$$\frac{dx_i}{d\lambda_c} = -\frac{x_c}{z_c} \frac{z_0}{\mu \lambda_c}$$

and

$$\frac{dz_i}{d\lambda_c} = -\frac{\frac{z_0}{\mu}}{\lambda_c}. \tag{5.10}$$

If the mean wavelengths during recording and reconstruction are nearly the same ($\lambda_c \approx \lambda_r$), then $\mu = 1$. The bandwidth of the light source is $d\lambda_c$.

As an example for the application of Eq. (5.10) a high-pressure Hg lamp with a mean wavelength of $\lambda_c = 546$ nm and a bandwidth of $d\lambda_c = 5$ nm is assumed. The hologram is created using a He–Ne laser ($\lambda_r = 636$ nm). The direction of the illumination during reconstruction is $30°$, i.e., $x_c/z_c = \tan 30°$. Using a distance of $z_0 = 5$ cm between the object and image and the hologram the resulting shift of the image is $dx_i = 0.3$ mm.

5.2.2
Image-Plane Holograms

For image-plane holograms the image is positioned in the plane of the hologram (Section 4), i.e., $z_0 = 0$. This case puts minimum requirements on the bandwidth of the light source for reconstruction. Even white light sources may be used since $dx_i = 0$. However, object points outside of the hologram plane show strong chromatic aberrations which can be calculated from Eq. (5.10). A similar statement applies to the size of the light source; for $z_o = 0$ it has no influence on the sharpness of the image.

5.3
Image Luminance

The brightness of the reconstructed images depends primarily on the diffraction efficiency of the hologram. It is defined as the ratio of the diffracted light and the incident light of the image. A detailed description of this topic will be given in Chapter 6. Besides the diffraction efficiency the brightness of the image is influenced by geometric factors during recording and reconstruction.

5.3.1
Without Pupil

An off-axis hologram is investigated according to Fig. 5.2a. The virtual image at distance z_i is viewed by an observer at distance D. In doing so it is assumed that the intensity distribution in the image and hologram is relatively homogeneous. The hologram of area A_H is illuminated with monochromatic light of intensity I. The diffraction efficiency is ϵ. Thus the diffracted light power of the hologram is

$$P = \epsilon I A_H. \tag{5.11}$$

In illumination technologies the luminance of an emitting area is defined as light flux per unit area and emitting solid angle. These physiological units are seldom used in laser technologies. Instead the term "radiation density" is introduced; it defines the emitted power per unit area and solid angle [4]. The solid angle which is illuminated by the image is given by $\Omega_H = A_H/z_i^2$, see Fig. 5.2a. Hence the radiation density of the image becomes

$$L = \frac{\epsilon I A_H}{\Omega_H A_i}$$
$$= \frac{\epsilon I z_i^2}{A_i}. \tag{5.12}$$

Accordingly the radiation density rises with increasing distance of the object from the hologram. The rise in brightness follows from the fact that the illumination covers a smaller solid angle with the disadvantage of a smaller viewing angle.

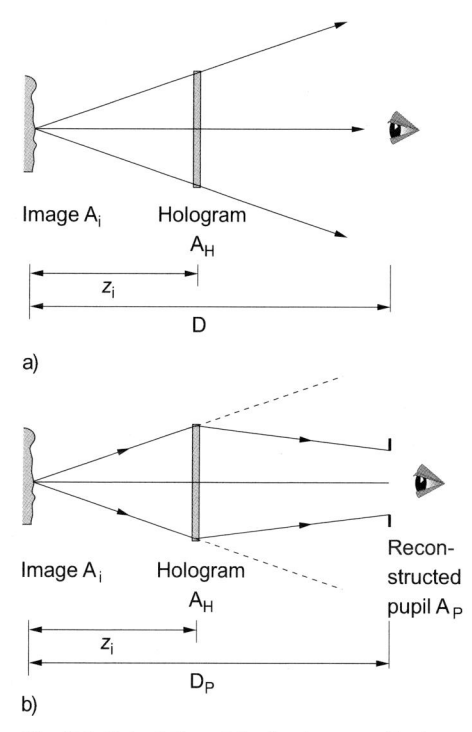

Fig. 5.2 Calculation of the luminance of holographic images: (a) off-axis hologram without pupil and (b) hologram with pupil.

5.3.2
With Pupil

If a hologram of a real image is created which is formed by a lens or another hologram, also the pupils or other apertures are imaged. One example is the rainbow holograms according to Fig. 4.4 where the aperture slit lies in front of the hologram as an image. The general case of a hologram with a pupil is shown in Fig. 5.2b.

The solid angle in which the image irradiates is given by $\Omega = A_P/D_P^2$ where A_P is the area and D_P is the pupils' distance from the image. Again it is assumed that the image radiates homogeneously. Instead of Eq. (5.12) the resulting radiation density is

$$L = \frac{\epsilon I A_H}{\Omega_P A_i}. \tag{5.13}$$

Compared to a hologram without pupil the radiation density is increased by a factor of Ω_H/Ω_P. This explains the brilliance of rainbow holograms. Even without any calculation it becomes clear that the brightness rises if the light is concentrated on a smaller area or a smaller solid angle. A disadvantage is still the limitation in viewing angle.

5.3.3
Image-Plane Holograms

Only minor demands are put onto the spectral purity and the size of the light source if the image lies in the hologram plane as in image-plane holograms (see Section 5.2). An adverse effect is that only those parts of the hologram deliver information which are of the same size as the image, i.e., $A_H = A_i$. This simplifies Eq. (5.13) to

$$L = \frac{\epsilon I}{\Omega_P}. \tag{5.14}$$

The radiation intensity is independent of the size of the hologram. The radiation density is reduced because only that part of the hologram contributes to the diffraction which lies in the area of the image. If the object is slightly moved away from the hologram plane the information is recorded in the complete hologram layer. The brightness rises but also do aberrations induced by the extension and spectral bandwidth of the light source.

5.4
Speckles

If a diffuse reflecting object is illuminated with laser light it is diffracted by discrete microscopical units. The diffracted light waves are coherent and interfere with each other. The interference creates bright and dark spots, the so-called speckles. The size of the granulation is proportional to the smallest resolvable distance that is achieved by an optical system such as, e.g., the eye, a camera or a hologram plate (see Section 15.6). The larger the aperture the smaller the resolvable distance gets and therefore the size of the speckles. The appearance of speckles is a serious problem in holography. In the following some techniques will be described which can be used to reduce the influence of speckles.

5.4.1
Diffuser

Speckles can be avoided for two-dimensional transparent objects if the resolution of the optical system is high enough. For holography this means that the plate has to be large and the object needs to be close by. But for such objects diffraction effects from dust particles and scratches are observable which appear very disturbing in the reconstructed image. This effect can be diminished by diffuse illumination of the object using a diffusing screen. The diffuser itself can be the cause of speckles if its structure is not resolved by the holographic system. Speckles caused by the diffuser can be avoided if the diffraction angle is small enough such that the entire light can pass the aperture of the imaging system.

Fourier holograms do have particular advantages since interference patterns caused by dust are not recorded, see Fig. 4.6. Although it is unfavorable that most of the object wave is recorded in a small area of the hologram, this causes large fluctuations in intensity, which leads to nonlinearities (Section 4.5). This can again be avoided by the use of a diffusing screen which expands the object wave. Hence the information of an object point is distributed over the whole hologram plate which has the effect that if the hologram is broken into pieces the complete image is retained in each fragment although with reduced resolution.

Diffusers are, e.g., used when creating holograms of persons since it prevents focusing of the laser beam onto the retina of the eye.

5.4.2
Resolution

It was already mentioned that the speckle size is reduced in optical systems with high resolution. This is also true for three-dimensional objects and is achieved by the hologram plate being large and the object being close by. By doing so the speckle size can become smaller than the granulation of the holographic layer. The recording averages over several speckles, which drastically reduces the intensity variations. Similar holds if the speckles on the retina become smaller than the distance of the photoreceptors.

5.4.3
Incoherent Illumination

Speckles are caused by the coherence of the laser light. During reconstruction of holograms the speckles can be reduced by decreasing the coherence of the illuminating light, e.g., by the use of tungsten filament lamps with small coils or similar emitters. In white-light holograms speckles are usually not a problem. The spatial coherence can also be reduced by using moving diffusers.

5.4.4
Further Techniques

A different technique for the reduction of speckles is to record multiple holograms of the same object with different positions of the diffusing screen. The speckles average out when the reconstructed images are superposed. Another technique samples a large hologram using a movable aperture. Thereby the speckles move within the image and can be reduced by temporal averaging.

5.5
Resolution

Optical devices, e.g., microscopes, have a limited resolution which is governed by diffraction effects at the edge of the lens. Imaging a single point results in a diffraction disc as an image. The diameter of this disc represents a distance d [1] on the object:

$$d = \frac{2z_0 \cdot \lambda}{D}, \tag{5.15}$$

where z_o is the distance of the object, λ is the wavelength, and D is the diameter of the lens. The smallest distance of points that can be resolved on an object is given by the distance d. The larger the lens diameter, the better the resolution gets.

A hologram can also be thought of as an imaging optical element. Therefore it is not astonishing that the resolution of a hologram is also approximately described by Eq. (5.15). Then D is the diameter of the hologram and z_o is the distance of the object from the hologram. If the information of a point is not recorded on the whole hologram then D is the diameter of the effective recording area. The validity of Eq. (5.15) for holography requires that the recording layer is able to record sufficiently high spatial frequencies.

Problems

Problem 5.1 Verify Eq. (5.2). What is the relationship between z_i and z_o for the plane reference and reconstruction waves and $m = 1$ and $\mu = 1$? Show that the last result can be derived easily from Eq. (5.3).

Problem 5.2 Calculate the lateral magnification of an object point $P(x_o, y_o, z_o)$ for plane reference and reconstruction waves ($m = 1, \mu = 1$). Are there other reference and reconstruction waves to get the same lateral magnification?

Problem 5.3 Verify Eq. (5.5b) using Eq. (5.5a).

Problem 5.4 Calculate the ratio of radiances of a rainbow hologram and a full aperture hologram. Assume that the distance of H1 (master hologram) from H2 (rainbow or full aperture hologram) is about 15 cm, the holograms H1 and H2 have a diameter of 10 cm, the object and the image of the object have a diameter of 5 cm, the slit to exposure the rainbow hologram is 5 mm.

Problem 5.5 Calculate the diameter of the diffraction disk of an object point from Eq. (5.15) for a hologram of about 10 cm width, an object distance of 10 cm from the holographic plate and a wavelength of 600 nm.

6
Types of Holograms

6.1
Introduction

6.1.1
Transmission and Reflection Holograms

In this chapter the properties of different hologram types are presented. One differentiates between transmission and reflection holograms depending on whether the hologram is to be viewed in transmitted or in reflected light. The geometrical setup during recording specifies which type of hologram is realized.

It was referred several times to that a hologram can be understood as a complicated diffraction grating. After development the grating is formed by the opaque silver grains. During reconstruction the light wave is diffracted and partially absorbed; hence these holograms are called "amplitude holograms." By using "bleaching baths" the silver can be converted into translucent halide or even be removed completely from the emulsion. The diffraction grating is then formed by areas of different index of refraction; a "phase hologram" is created.

6.1.2
Thick and Thin Holograms

Another characteristic to distinguish holograms is the thickness of the emulsion, d, compared to the mean lattice constant in the hologram, d_g. If

$$d \ll d_g,$$

one speaks of a "thin" hologram. For the case of

$$d \gg d_g,$$

one speaks of a "volume hologram."

Holography: A Practical Approach. Gerhard K. Ackermann and Jürgen Eichler
Copyright © 2007 WILEY-VCH Verlag GmbH & Co. KGaA, Weinheim
ISBN: 978-3-527-40663-0

This difference plays an important role in the diffraction efficiency, i.e., the brightness of the reconstructed image. Thin holograms in principle have a low diffraction efficiency whilst volume holograms exhibit a larger brightness in the reconstructed image. In this chapter equations for the diffraction efficiency of the individual holograms are derived and discussed.

6.2
Thin Holograms

The determination of the diffraction efficiency is described in detail in the literature ([8] to [9]). For simplification it is assumed that two plane waves create a grating by interference inside the photographic layer.

6.2.1
Thin Amplitude Holograms

The plane waves create structures by interference inside the photographic plate which are constant in the y-direction (Fig. 6.1). The amplitude transmission $t(x)$ is represented in Section 2.3 by a cosine function with the spatial frequency σ:

$$t(x) = \bar{t} + t_1 \cos(2\pi\sigma x), \tag{6.1}$$

where $\bar{t}$ represents the mean transmission amplitude and t_1 the modulation depth of the grating. In the following the maximal diffraction efficiency is to be derived. The mean transmission has to be between 0 and 1 and for this regular grating is assumed to be $\bar{t} = 0.5$. Half of the light amplitude is lost inside the grating structure by absorption. Equation (6.1) yields for $t_1 = 0.5$:

$$t(x) = 0.5 + 0.5 \cos(2\pi\sigma x). \tag{6.2}$$

With the help of the Euler relation

$$\cos(2\pi\sigma x) = 0.5(e^{i2\pi\sigma x} + e^{-i2\pi\sigma x})$$

the result for Eq. (6.2) is

$$t(x) = 0.5 + 0.25(e^{i2\pi\sigma x} + e^{-i2\pi\sigma x}). \tag{6.3}$$

During reconstruction the amplitude of the incoming wave is modulated with $t(x)$. The last two terms in Eq. (6.3) represent two waves with a relative amplitude of 1/4. The intensity of both waves is given by the square of the corresponding amplitudes. The ratio of the intensities of the diffracted waves in the first diffraction orders I_{+1} and I_{-1} and the incoming intensity is called

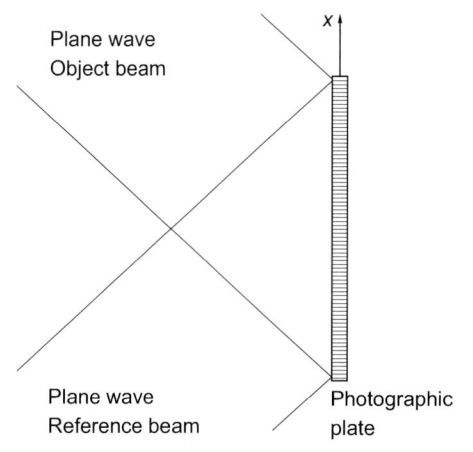

Fig. 6.1 Principle of a holographic recording with two plane waves. A carthesic coordinate system is indicated. The *y*-axis lies in the plane of the plate and is perpendicular to the shown *x*-axis.

"diffraction efficiency ϵ:"

$$\epsilon = \frac{I_{+1}}{I} = \frac{I_{-1}}{I} = \frac{1}{16} = 6.25\%.$$

Maximal 6.25% of the incoming light is diffracted in the first grating orders. Hence the intensity of the object waves is very weak.

6.2.2
Thin Phase Holograms

Using bleaching techniques which are described in detail in Chapter 14 it is possible to convert an amplitude modulation into a phase modulation. Thus the loss of light in the illuminated areas of the hologram due to the developed silver can be avoided. These silver grains are removed from the photographic layer by the solving bleach. Two emulsion areas with different densities are created containing either unexposed silver halide or just the gelatin layer. The index of refraction as well as the running time of the light is different for both the exposed and the unexposed areas. The phase of the light is periodically modulated after passing the grating. In general the complex transmission $\mathbf{t}(x)$ is written as

$$\mathbf{t}(x) = t(x)e^{i\Phi(x)}. \tag{6.4}$$

In amplitude holograms the amplitude $t(x)$ is modulated and $\Phi(x)$ is constant (Eq. (6.1)). In phase holograms it behaves exactly reverse. The phase of a periodic grating structure $\Phi(x)$ is represented by a cosine function:

$$\Phi(x) = \Phi_0 + \Phi_1 \cos(2\pi\sigma x). \tag{6.5}$$

For pure phase holograms $t(x) = 1$ can be set:

$$t(x) = e^{i\Phi_0} e^{i[\Phi_1 \cdot \cos(2\pi\sigma x)]}. \tag{6.6}$$

For weakly modulated gratings Φ_1 is small and the exponential function can be expanded into a series. In the first approximation this results in

$$t(x) = e^{i\Phi_0}(1 + i\Phi_1 \cos[2\pi\sigma x]).$$

The factor $e^{i\phi_0}$ is of no relevance for this consideration. The right-hand side of this equation has a structure similar to Eq.(6.1). In thin phase gratings with weak modulation like in amplitude gratings only the first diffraction orders appear. The diffraction efficiency is

$$\epsilon = \left(\frac{\Phi_1^2}{4}\right) \ll 1.$$

The above expansion into a Taylor series is not valid for strong modulations. Applying the Euler relation to Eq. (6.6) it follows that

$$e^{i\Phi_1 \cos(2\pi\sigma x)} = \cos[\Phi_1 \cos(2\pi\sigma x)] + i \sin[\Phi_1 \cos(2\pi\sigma x)].$$

Smith [10] has given an expansion into Bessel functions for this. It states

$$\cos(\Phi_1 \cos(2\pi\sigma x)) = J_0(\Phi_1) + 2\sum_{n=1}^{\infty}(-1)^n J_{2n}(\Phi_1) \cos(2n(2\pi\sigma x))$$

and

$$\sin(\Phi_1 \cos(2\pi\sigma x)) = 2\sum_{n=0}^{\infty}(-1)^{n+2} J_{2n+1}(\Phi_1) \cos((2n+1)(2\pi\sigma x)).$$

Thus it follows

$$\begin{aligned}t(x) = J_0(\Phi_1) + 2iJ_1(\Phi_1) \cos(2\pi\sigma x) - 2J_2(\Phi_1) \cos(2(2\pi\sigma x)) \\ -2iJ_3(\Phi_1) \cos(3(2\pi\sigma x)) + \cdots\end{aligned} \tag{6.7}$$

Equation (6.7) shows that several diffraction orders can be expected whose observable intensity is proportional to the squares of the Bessel functions. Figure 6.2 shows the Bessel functions J_i^2 for $i = 0, 1, 2, 3$. Taking into account only the first two summands of Eq. (6.7) (6.7) this yields

$$t(x) = J_0(\Phi_1) + 2iJ_1(\Phi_1) \cos(2\pi\sigma x).$$

Using the Euler relation this can be written as the sum of two exponential functions:

$$t(x) = J_0(\Phi_1) + iJ_1(\Phi_1)(e^{i2\pi\sigma x} + e^{-i2\pi\sigma x}). \tag{6.8}$$

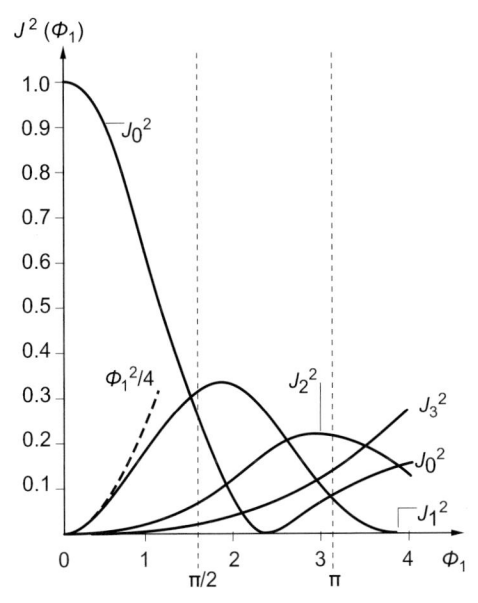

Fig. 6.2 The diffraction efficiency of thin phase holograms is given by the squares of the Bessel functions J_i^2 which are shown in the figure for the diffraction orders $i = 0, 1, 2,$ and 3. Φ_1 represents the variation in phase within the grating (Eq. (6.5)).

The expressions in the brackets of Eq. (6.8) represent the virtual and the real images. From Fig. 6.2 the maximal value of J_1^2 can be extracted:

$$J_1^2 = 0.339.$$

Thus the resulting maximal diffraction efficiency ϵ_{max} is

$$\epsilon_{max} = 0.339 = 33.9\%.$$

The diffraction efficiency is hence much larger than that in amplitude holograms although still much lower than 100%. Equation (6.7) shows that in principle not only the two images of the first order are observed. If Φ_1 is large enough also images of higher order appear. The statement, often found in textbooks, that sinus gratings only create one diffraction order is not true for thin phase holograms with strong modulation. In most of the realized experiments however only one diffraction order appears which is because J_2 is too small.

6.3
Volume Holograms

The first recording of a volume hologram was performed by Denisyuk [11] in 1962. Contrary to thin holograms the thickness of the emulsion does play a role in volume holograms. The important parameter for the classification is the ratio of the grating constant or the spatial frequency and the thickness of the light sensitive layer. In Fig. 6.3 examples for thick transmission and reflection holograms are shown. The indicated bars represent grating planes which are perpendicular to the plane of paper. The position of the particular planes which act as partially transparent mirrors is given by the angle between reference and object wave during recording.

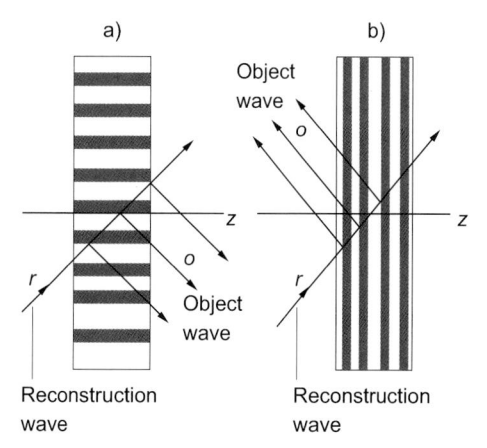

Fig. 6.3 Principle representation of a volume transmission hologram (a) and a reflection hologram (b) for the determination of the diffraction efficiency [12].

The outlined transmission hologram in Fig. 6.3 is generated when two plane waves pass the light sensitive layer symmetrically from one side (a). An exposure from opposite sides creates a reflection hologram (b). In the figure the reconstruction of the object wave by the reference wave is shown.

6.3.1
Theory of Coupled Waves

The theory for the calculation of the diffraction efficiency was developed by Kogelnik [9]. Only an abstract will be described here (see also [3]). Only grating planes will be considered here that are perpendicular or parallel to the surface of the emulsion. Accordingly only plane waves for the object and reconstruction waves will be considered.

The structure of a volume hologram with its planes of developed and not developed silver reminds of a crystal lattice. Similar to the diffraction of x-rays

6.3 Volume Holograms

at the lattice planes the position of the diffraction orders is given by the Bragg condition (see Appendix B). Diffraction only occurs if the reconstruction wave impinges at a certain angle. If the distance of the lattice planes is d_g then the result for this Bragg angle ϑ is

$$2d_g \sin \frac{\vartheta}{2} = \lambda, \tag{6.9}$$

where ϑ is the angle of the reference wave during recording. An object wave **o** only appears if the reconstruction wave impinges at the angle ϑ. Deviations from the Bragg angle are not considered.

The diffraction efficiency is calculated by calculating the amplitude of the diffracted wave. Kogelnik assumes the classical wave equation. The reconstruction wave **r** is attenuated when passing the hologram whilst the object wave **o** is amplified (Fig 6.3). Therefore, a coupling between these two waves must exist; the amplitudes depend on the depth z of the hologram.

The complex amplitudes of the waves **r** and **o** are given by

$$\mathbf{r} = r(z)e^{-i\mathbf{k}_r \mathbf{x}} \tag{6.10}$$

and

$$\mathbf{o} = o(z)e^{-i\mathbf{k}_o \mathbf{x}}.$$

$\mathbf{k}_r$ and $\mathbf{k}_o$ are connected by the following vector equation:

$$\mathbf{k}_o = \mathbf{k}_r - \boldsymbol{\sigma}, \tag{6.11}$$

where $\mathbf{k}_r$ and $\mathbf{k}_o$ are the wave vectors of the two waves. They have the same length:

$$|\mathbf{k}_r| = |\mathbf{k}_o| = n\left(\frac{2\pi}{\lambda}\right),$$

where n is the refraction index of the emulsion. $\boldsymbol{\sigma}$ describes the lattice vector which is perpendicular to the lattice planes having the distance d_g:

$$|\boldsymbol{\sigma}| = \frac{2\pi}{d_g}.$$

From Fig. 6.4 it can be seen that Eq. (6.11) represents the BBragg condition. The sum of the two waves

$$\mathbf{E} = \mathbf{r} + \mathbf{o}$$

has to satisfy the classical wave equation:

$$\Delta \mathbf{E} + k^2 \mathbf{E} = 0,$$

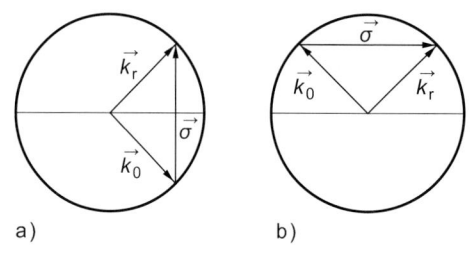

Fig. 6.4 Vectorial representation of the Bragg condition for the two cases shown in Fig. 6.3.

where k is the absolute value of the wave vector in the medium. In vacuum the following applies:

$$K = \frac{2\pi}{\lambda}.$$

In a medium the complex refraction index has to be considered:

$$\mathbf{n} = n(1 - i\alpha),$$

and it is

$$\mathbf{k} = K\mathbf{n} \quad \text{with} \quad k = |\mathbf{k}|.$$

For the distribution of the refraction index n and the absorption coefficient α a sinusoidal distribution is assumed:

$$n = n_0 + n_1 \cos(\sigma x)$$
$$\alpha = \alpha_0 + \alpha_1 \cos(\sigma x). \tag{6.12}$$

$\mathbf{x}$ denotes the vector in the x-direction. It applies approximately

$$k^2 = n_0^2 K^2 - 2i\alpha_0 n_0 K + 4\kappa n_0 K \cos(\sigma x) \tag{6.13}$$

with

$$\kappa = \frac{\pi n_1}{\lambda} - i\frac{\alpha_1}{2}. \tag{6.14}$$

κ, the coupling constant, describes the energy transfer from the object to reference wave. κ becomes zero for a homogeneous medium ($n_1 = \alpha_1 = 0$).

Putting Eqs. (6.10) and (6.13) into the wave equation coupled differential equations result. Summarizing all terms with corresponding e-functions according to Eq. (6.10) yields

$$\mathbf{r}'' - 2i\mathbf{r}'k_{r,z} - 2i\alpha n_0 K\mathbf{r} + 2\kappa n_0 K\mathbf{o} = 0$$

and

$$\mathbf{o}'' - 2i\mathbf{o}'k_{r,z} - 2i\alpha n_0 K\mathbf{o} + 2\kappa n_0 K\mathbf{r} = 0.$$

The terms having one or two primes represent the first and second derivative with respect to the spatial coordinate z, respectively. Assuming that the energy transfer between the coupled waves is so slow that $\mathbf{r}'$ and $\mathbf{o}'$ only change weakly, $\mathbf{r}''$ and $\mathbf{o}''$ can be neglected [12]. Using

$$k_{r,z} = k_r \cos \vartheta = n_0 K \cos \vartheta$$

and

$$k_{o,z} = k_o \cos \vartheta = n_0 K \cos \vartheta$$

one can write

$$\mathbf{r}' \cos \vartheta + \alpha \mathbf{r} = -i\kappa \mathbf{o} \tag{6.15}$$

and

$$\mathbf{o}' \cos \vartheta + \alpha \mathbf{o} = -i\kappa \mathbf{r}.$$

Here ϑ is the Bragg angle. The equations are easy to interpret. The changes in both waves are due to the transfer to the other wave and due to the absorption. If the coupling constant was $\kappa = 0$ then $\mathbf{r}$ would only be weakened by α_0, i.e., the general absorption by passing through the photographic layer.

6.3.2
Phase Holograms

For a phase hologram with negligible absorption ($\alpha_0 = \alpha_1 = 0$) there is a solution for the diffracted wave at point $z = d$, i.e., directly when leaving the photographic layer:

$$\mathbf{o}(d) = -i \sin \Phi \tag{6.16}$$

with the modulation parameter

$$\Phi = \frac{\pi n_1}{\lambda} \frac{d}{\cos \vartheta}. \tag{6.17}$$

The parameter Φ depends on the variation of the refraction index n_1, the layer thickness d, the Bragg angle ϑ, and the wavelength λ (Eq. (6.1)). The diffraction efficiency ϵ for transmission holograms is proportional to the square of the amplitude which is contained in Eq. (6.16):

$$\epsilon = \left| \mathbf{o}(d)^2 \right| = \sin^2 \Phi. \tag{6.18}$$

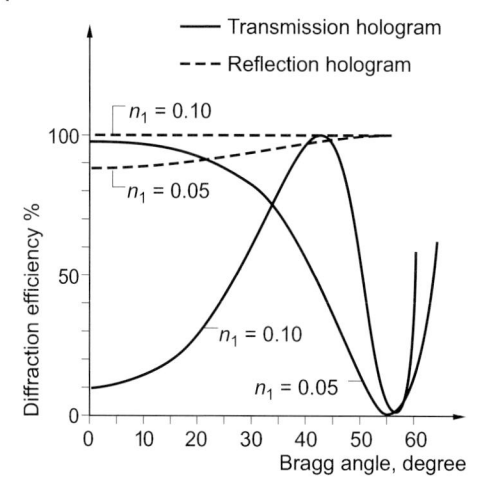

Fig. 6.5 Diffraction efficiency ϵ for phase holograms as a function of the Bragg angle and a variation of the refraction index n_1. The basis for the figure is Eqs. (6.18) (transmission holograms) and Eq. (6.19) (reflection holograms).

Here the amplitude of the reconstruction wave was tacitly assumed to be 1. The diffraction efficiency reaches 100% ($\epsilon = 1$) for $\Phi = \pi/2$. If $\Phi > (\pi/2)$ the diffraction efficiency ϵ decreases again.

Assume the following example: if $\vartheta \approx 45°$, $d \approx 7$ µm and $\lambda = 633$ nm the result is $\Phi \approx n_1 \cdot 50$. For a variation in refraction index of $n_1 = 0.03$ this yields $\Phi = 1.5 \approx \pi/2$ ($\epsilon = 1$).

If the absorption in the hologram cannot be neglected, a damping factor α_0 has to be introduced in Eq. (6.16):

$$\mathbf{o}(d) = -i e^{\frac{-\alpha_0 d}{\cos \vartheta}} \sin \Phi,$$

which results in a general decrease of the diffraction efficiency.

Equation (6.15) for reflection holograms results in a slightly different equation for the diffraction efficiency:

$$\epsilon = \tanh^2(\Phi). \tag{6.19}$$

The diffraction efficiency continuously increases up to $\epsilon = 1$. This value is quickly reached for $\Phi = \pi/2$ (Fig. 6.5).

6.3.3
Amplitude Holograms

The refraction index for an amplitude hologram is $n_1 = 0$. The coupling constant only contains the parameter α_1. The following expression applies to a transmission hologram at the Bragg angle ϑ:

$$\mathbf{o}(d) = -e^{-\frac{\alpha_0 d}{\cos \vartheta}} \sinh \Phi_a. \tag{6.20}$$

Equation (6.20) takes into account the above-mentioned damping factor because the recording of the information is due to the modulation in the absorption (α_0, α_1). The term Φ_a describes the modulation:

$$\Phi_a = \frac{\alpha_1 \cdot d}{2 \cdot \cos \vartheta} \tag{6.21}$$

The largest value for the diffraction efficiency ϵ is reached for $\alpha_0 = \alpha_1$ and $(\alpha_1 d)/(\cos \vartheta) = \ln 3$:

$$\epsilon_{max} = 0.037 = 3.7\%$$

The calculation for the object wave $\mathbf{o}$ of reflection amplitude holograms leads to the hyperbolic cotangent function. The calculation is not described in detail because of the small meaning that amplitude holograms have due to their small diffraction efficiency. The maximal diffraction efficiency is

$$\epsilon = 0.072 = 7.2\%$$

6.3.4
Comparison of Diffraction Efficiency

The diffraction efficiency is shown in Fig. 6.5 as a function of the Bragg angle ϑ. The variation of the refraction index n_1 was chosen as a parameter. The results clarify that the diffraction efficiency for reflection holograms follows a monotonic growing function ($\epsilon = \tanh^2 \Phi$). The value of ϵ for $\vartheta = 0$ depends on the parameter n_1. The parameters chosen here lead to large starting values of the diffraction efficiency.

For transmission holograms (Eq. (6.18)) ϵ follows a periodic function. As shown in the examples in the figure the diffraction efficiency for small angles ϑ can be small as well as very large. If $d = 7$ µm and $\lambda = 633$ nm it follows that for $n_1 = 0.1$ and $\vartheta = 0$ the modulation parameter is $\Phi \approx 1.1\pi$. According to Eq. (6.18) ϵ is very small and increases with ϑ. If $n_1 = 0.05$ is chosen with all other parameters unchanged the result is $\Phi \approx 0.55\pi$; ϵ is almost equal to 1.0 and decreases with ϑ. For large Bragg angles the function for the diffraction efficiency shows strong oscillations making it difficult to state the value that can be expected.

The question about the hologram being thick or thin is not easily answered. Theoretically for thick holograms only one diffraction order should be observable. Investigations however show that also for grating constants much lower than the thickness of the photosensitive layer two or more diffraction orders appear.

The approximation described here is only valid for small values of n_1 and α_1. In Eqs. (6.13) and (6.14) only those terms that are linear in n_1 and α_1 were considered. For large values of n_1 and α_1 this is not valid. From the theory of diffraction gratings [3, 27] it is known that under this condition higher orders appear and the observable intensities are proportional to the square of the Bessel function.

6.3.5
Distinction Criteria for Holograms

Klein and Cook [13] have developed an exact criterion for the distinction between thin and volume holograms. They define a parameter Q:

$$Q = \frac{2\pi\lambda_0 d}{n_0 d_g^2}. \tag{6.22}$$

The fundamental parameter in Eq. (6.22) is the ratio of the emulsion thickness d and the grating constant d_g. Therefore Q is a measure to distinguish between types of holograms. For $Q \ll 1$ it is a thin hologram, for $Q \gg 1$ it is a volume hologram. Also important is the additionally defined modulation parameter Φ introduced in Eq. (6.17); only for small values can the Q-parameter be used as the only distinction criterion. With increasing Φ a large transition region appears where the Q-parameter does not deliver unambiguous statements. The more stringent condition for a thin hologram not only includes a low value for Q but the additional condition:

$$Q'\Phi < 1, \tag{6.23}$$

where $Q' = \frac{Q}{\cos\vartheta}$. For a volume hologram,

$$\frac{Q'}{\Phi} > 20. \tag{6.24}$$

Most of the realized results in laboratories fall within the excluded transition region of inequalities (6.23) and (6.24) and are therefore not unambiguously classifiable as thin or volume holograms.

If the variations in intensity become too large, the nonlinear region of the exposure curve is reached. The $\cos^2$-function of the intensity variation can no longer be registered as a $\cos^2$-function in the photosensitive layer. The distribution of the recorded grating is more like a rectangular function. Due to these disturbances higher diffraction orders appear. Similar holds for phase holograms.

Summarizing the above considerations it becomes clear that the diffraction efficiency of amplitude holograms has tight limits. On the other hand, the diffraction efficiency of phase holograms can reach 100% under optimal conditions. The question for the number of diffraction orders can be answered unambiguously for thin or volume holograms if the modulation parameter is known. However, the results very often fall within the described transition region which hampers a distinct statement.

Problems

Problem 6.1 Thin amplitude holograms show a diffraction efficiency of about 6%. Verify that this is true for the real image. Make the assumption that the mean transmission of the hologram is 0.5.

Problem 6.2 Show that the diffraction efficiency for thin phase holograms is

$$\varepsilon = \frac{\Phi_1^2}{4}.$$

Problem 6.3 Give a physical explanation of the three terms of the coupled wave equation (6.15).

Problem 6.4 Solve Eq. (6.15) and find solutions for the object wave $o(z)$; assuming negligible absorption. The solution has to fulfill the boundary conditions: $r(0) = 1$ and $o(0) = 0$.

Problem 6.5 Give the vector representation of the Bragg condition for a reflection hologram with a plane object wave impinging the holographic plate perpendicular to the surface and a plane reference beam from the other side under an angle of 60° with respect to the surface of the holographic plate (Fig. 6.4). Derive a formula for the spatial frequency.

Problem 6.6 A hologram shows a diffraction efficiency of 50%. Assumed no absorption, thickness of emulsion layer 7 µm, Bragg angle $\vartheta = 30°$ and $\lambda = 632.8$ nm, calculate the diffraction index variation n_1.

Problem 6.7 Calculate the diameter of the diffraction disk of an object point from Eq. (5.15) for a hologram of about 10 cm width, an object distance of 10 cm from the holographic plate, and a wavelength of 600 nm.

Part 2 Basic Experiments

7
Optical Systems and Lasers for Holography

7.1
Coherence and Interferometers

7.1.1
Coherence

The term "coherence" means "connection" and describes how closely a real wave field resembles the statistically varying amplitude and phase of an ideal wave. Ideal waves with exactly defined amplitude and phase are called "coherent." Conventional light sources and also lasers emit waves that resemble ideal waves only in a very small space–time domain. Therefore they are referred to as "partially coherent."

The coherence properties of light are especially important for setups using interference like, e.g., in holography. With coherent light interference effects can be observed, i.e., constructive and destructive superposition of the field amplitudes of light waves. On the other hand, no interferences appear for incoherent light; the intensities superpose additively and the creation of a hologram is not possible. For partially coherent light the contrast of the interference effects is decreased.

7.1.2
Spatial Coherence

It is distinguished between spatial and temporal coherence. The spatial coherence describes the correlation of field strengths or amplitudes in two different points of a wave field at a given point in time. The investigation of the coherence can be done with an interference experiment using two pinholes in the wave field (Fig. 7.1). Complete coherence creates interference fringes that are modulated down to zero if the two amplitudes are the same. For partial coherence the contrast is decreased.

A laser oscillating in a single transversal mode, e.g., TEM_{00}, is completely spatially coherent. A transversal multimode laser has a low spatial coherence

---

Holography: A Practical Approach. Gerhard K. Ackermann and Jürgen Eichler
Copyright © 2007 WILEY-VCH Verlag GmbH & Co. KGaA, Weinheim
ISBN: 978-3-527-40663-0

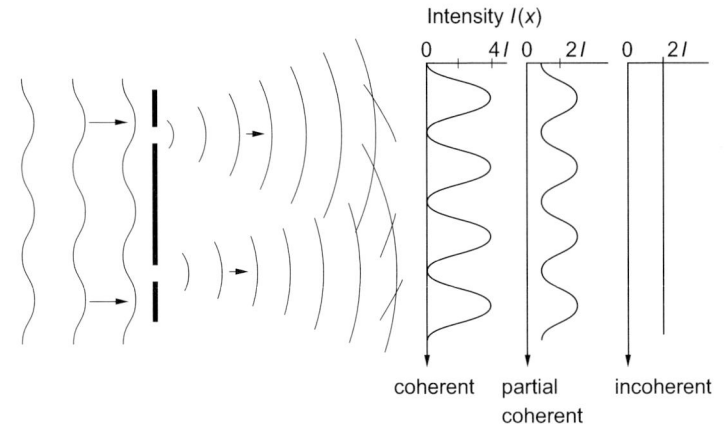

Fig. 7.1 Principle of measuring spatial coherence. Two points are defined using pinholes in the wave field. Behind these holes spherical waves are created which interfere in the observers plane. The contrast of the intensity distribution $I(x)$ is a measure for the degree of coherence. For incoherent light $I(x) = 2I$ and is constant. For incoherent light $I(x)$ varies between 0 and 4I.

because the transversal modes have different frequencies and therefore temporal varying phase differences. In holography mostly TEM_{00} lasers are used so that the radiation is spatially coherent.

7.1.3
Temporal Coherence

In contrast to their spatial coherence lasers used for holography are not ideal temporal coherent. For temporal coherence the field strengths or amplitudes of a light wave are compared or "correlated" at a fixed point in space but for different points in time. In general, it can be observed that the field strengths for two different points in time have a constant phase difference. But if the spatial difference exceeds a specific maximum value, the so-called "coherence time t_c," the phase difference varies statistically.

The coherence time can be measured experimentally with a Michelson interferometer (Fig. 7.2). A light beam is split into two separate beams using a beam splitter. These are reflected into themselves by one mirror each and are recombined by the beam splitter with a slight angular offset between the two traveling directions. In the overlapping area of the two single beams a system of interference fringes is created.

Both mirrors are positioned in different and variable distances to the beam splitter so that the beams are retarded against each other. The retardation can be chosen so large that no interference occurs anymore. Therefore coherence length and coherence time are measurable. The coherence length is defined as

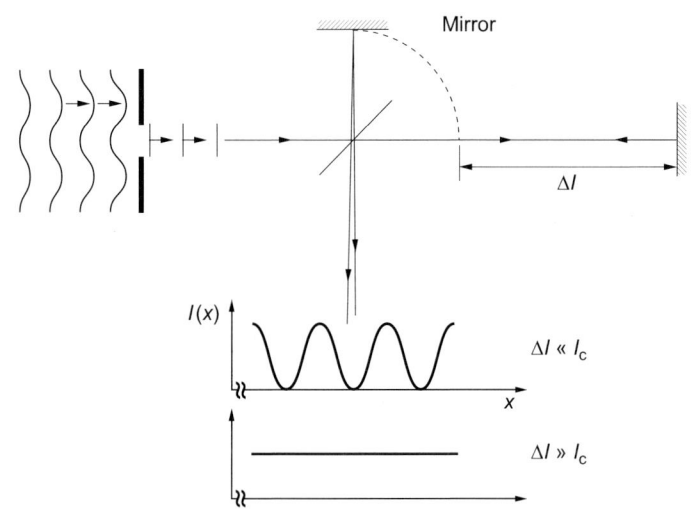

Fig. 7.2 Michelson interferometer to measure the coherence length l_c (temporal coherence). The intensity distribution $I(x)$ in the observer's plane depends on the path difference l of the two beams which are created by the beam splitter. The coherence length is the path difference for which the contrast decreases to $1/e = 37\%$.

the distance after which the contrast of the interference fringes has decreased to $(1/e) = 37\%$.

For conventional light sources the emission consists of single spontaneously emitted photons or wave packets with a duration τ which corresponds to the lifetime of the emitting energy level. From one wave packet to the other the phase varies statistically so that the following coherence time results:

$$t_c \approx \tau. \tag{7.1}$$

The duration of a wave packet and the lifetime τ are connected to the spectral width Δf of the wave by the following equation:

$$t_c \approx \frac{1}{\Delta f}. \tag{7.2}$$

Although lasers do not emit wave packets, but a wave with constant amplitude, Eq. (7.2) can also be applied to lasers. Thus the coherence length is limited by the bandwidth of the laser radiation.

The distance traveled by the light in the time t_c is called "coherence length:"

$$l_c = ct_c = \frac{c}{\Delta f}, \tag{7.3}$$

where c denotes the speed of light.

White light which contains the whole visible spectrum has a coherence length of about 1 μm. With good spectral lamps coherence lengths up to 1 m can be achieved but with very low intensity. For lasers the resulting coherence lengths reach, depending on the stabilization, from parts of a mm up to several km. For holography usually coherence lengths from 0.1 to 1 m are sufficient. This means that the path differences between the object and the reference wave must not exceed these values. The connection between axial modes and the coherence length of lasers is explained in the following section.

7.2
Modes and Coherence

Lasers for holography are operated in the basic transversal mode so that spatial coherence is achieved. The selection of this basic mode can be done very easily since higher transversal modes exhibit a larger beam diameter. By using a mode aperture with matching diameter inside the resonator higher modes can be suppressed.

7.2.1
Gaussian Beam

The basic mode of a laser is emitted as a so-called Gaussian beam [14]. The beam profile is given by a Gauss' function,

$$I(r) = I_{max} e^{-\frac{2r^2}{w^2}}, \tag{7.4}$$

where $I(r)$ is the radial intensity distribution. The beam radius w denotes the point in space where the intensity has decreased to $1/e^2 = 13\%$ with respect to the maximum value I_{max}. A diagram of the beam profile is shown in Fig. 7.3. It can be observed that a homogeneous illumination is difficult to achieve in holography. The beam diameter at the laser exit is given by the manufacturer. Inside this diameter is 86.5% of the laser power.

Laser beams do not propagate exactly parallel due to diffraction but with a certain divergence which is given by the smallest radius of the beam, w_0 (Fig. 7.9):

$$\Theta = \frac{\lambda}{\pi \cdot w_0}. \tag{7.5}$$

Here Θ represents half of the divergence angle and λ the wavelength of the radiation. Most of the time the smallest radius of the beam lies within the resonator and as an approximation the radius of the beam w_0 at the laser exit can be used in Eq. (7.5). For example, the resulting angle of divergence for a He–Ne laser with a beam diameter of 0.7 mm is $\Theta = 0.6$ mrad.

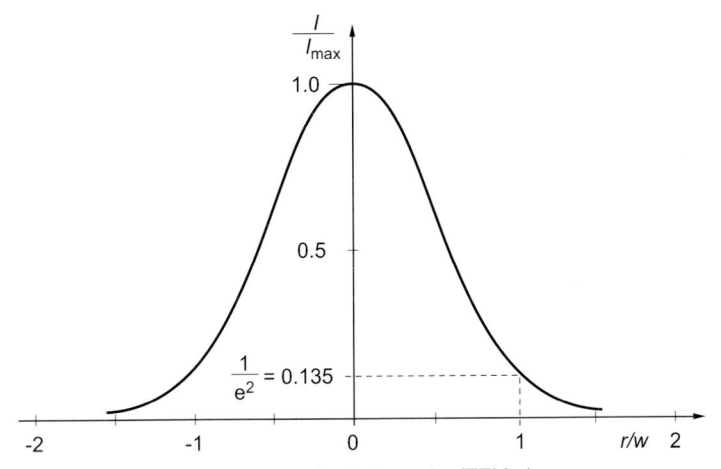

Fig. 7.3 Profile of a laser beam for holography (TEM$_{00}$).

7.2.2 Longitudinal Modes

The resonator of a laser consists of two mirrors at distance L. Standing light waves can be formed inside the system exhibiting discrete frequencies. The condition for a standing wave is that the resonator length L has to be a multiple m of half of the wavelength $\lambda/2$:

$$L = m\frac{\lambda}{2}. \tag{7.6}$$

With $f = c/\lambda$ the resulting frequencies and longitudinal modes of a laser are

$$f = \frac{m \cdot c}{2L}. \tag{7.7}$$

Hereby only frequencies appear that lie within the gain curve of the active medium. This is given by the line width and the losses inside the resonator. Figure 7.4 shows the modes of a laser and the gain curve. It is observable that for long resonators multiple longitudinal modes occur.

With short resonators it can be achieved that only one mode occurs. The bandwidth of a mode depends on the losses and the stability of the resonator; for a He–Ne laser, e.g., this can be around 3 Mhz. According to Eq. (7.3) with this value a coherence length of $l_c = 100$ m is achieved. According to Fig. 7.4 the condition for a resonator to be single moded is as follows:

$$\frac{c}{2L} \leq \Delta f_l, \tag{7.8}$$

where Δf_l is the width of the gain profile. For a He–Ne laser with $\Delta f_l \approx 1.5$ GHz the resulting maximum length for single-mode operation is $L \approx 10$ cm.

For an argon laser with $\Delta f_l \approx 6$ GHz the corresponding value is smaller by a factor of 4. A laser with such short lengths does not deliver enough power for holography.

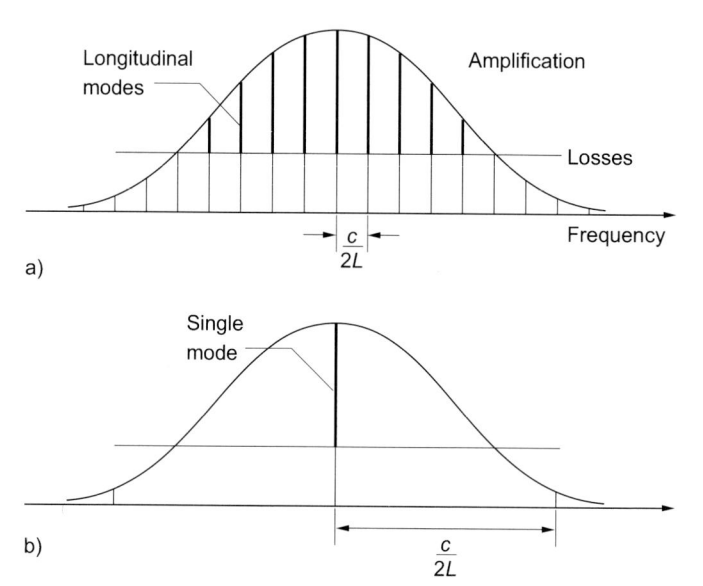

Fig. 7.4 Modes of a laser with length $c/2L$ and gain of the active medium: (a) long resonator and (b) short resonator (for $c/2L \leq \Delta f_l$ only one mode occurs).

7.2.3
Coherence Length

For long lasers according to Fig. 7.4 always several modes appear [15]. The number N can be easily estimated by dividing the whole line width $2\Delta f_l$ by the distance of the modes $c/2L$:

$$N \approx \frac{2\Delta f_l}{(c/2L)}. \tag{7.9}$$

The coherence length is calculated according to Eqs. (7.3) and (7.9) to

$$l_c \approx \frac{2L}{N}. \tag{7.10}$$

It increases with the resonator length L and decreases with the number N of longitudinal modes. (Eq. (7.10) is not valid for $N = 1$ since the width of a single mode has to be used for the bandwidth in Eq. (7.3).)

7.2.4
Etalon

Lasers with long resonators usually emit several longitudinal modes so that their coherence length is quite small. The conversion to monomode operation can be done by inserting an etalon in the resonator. An etalon consists of two plane parallel partially transparent mirrors and represents a short resonator with a length in the region of mm to cm. Thus the laser consists of two resonators, a long one and a short one. The laser can only oscillate if the modes in both resonators have the same frequencies. By proper adjustment this can be realized for a single mode so that monomode operation is possible.

In laser technology the etalon is mostly realized by a plane parallel glass plate which is mirrored on both sides. The plate thickness d and the index of refraction n govern the mode separation in the etalon:

$$\Delta f_D = \frac{c}{2n \cdot d}. \tag{7.11}$$

The width of the mode δf is called the "finesse F:"

$$F = \frac{\Delta f_D}{\delta f} = \frac{\pi \sqrt{R}}{1 - R}. \tag{7.12}$$

The finesse is governed by the reflection coefficient R of the two (equal) mirrors of the etalon.

For monomode operation the thickness of the etalon, d, has to be chosen in a way that Δf_D is larger than half the width of the gain profile Δf_l, Fig. 7.5.

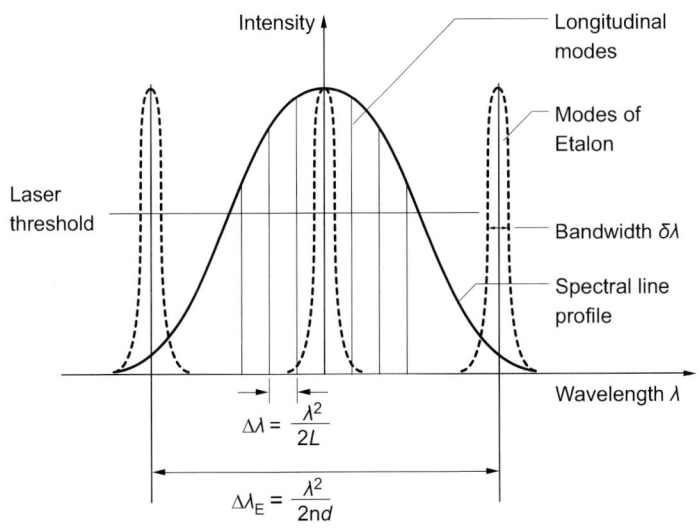

Fig. 7.5 Frequency selection (monomode) by using an etalon inside the laser resonator. One mode of the etalon and the resonator coincide at the maximum of the gain curve.

By slightly tilting the etalon one mode can be adjusted to the maximum of the gain curve. The length of the laser resonator is chosen such that the maximum output power is achieved and one laser mode is in the maximum of the gain profile. The finesse F is defined by fixing the reflection coefficient R of the etalon in a way, that the width δf is smaller than the longitudinal mode separation $c/\Delta L$. The widths of the gain profiles are known approximately for different types of lasers so that the necessary thickness of the etalon, d, and the reflection coefficient of the mirrors, R, to achieve monomode operation can be estimated.

7.3
Gas Lasers for Holography

The gas lasers used most often in holography are listed in Table 7.1. Those are the He–Ne laser and different ion lasers, especially argon, krypton, and He–Cd laser.

Tab. 7.1 Gas lasers for holography including power and wavelength

Laser	Wavelength		Power		Coherence length	
			optical	electrical	with	without etalon
	(μm)		(mW)	(kW)	(mm)	(m)
He-Ne	0.633	red	1 to 50	0.5	200	several m
He-Cd	0.442	violet	100	0.1	see He-Ne	
Ar+	0.458	bl.-viol.	200	7	mm	several m
	0.447	blue	400	"	"	"
	0.488	bl.-gr.	1000	"	"	"
	0.514	green	1500	"	"	"
Kr+	0.476	blue	50	"	"	"
	0.521	green	70	"	"	"
	0.647	red	500	"	"	"

7.3.1
He–Ne Laser

Most widely used is the relative inexpensive He–Ne laser with an output of about 10 mW. It is air cooled by convection and delivers a sufficient coherence length without etalon. The red color of the light (0.633 μm) is within the sensitivity region of holographic AgBr layers.

He–Ne lasers are offered with powers between 1 and 50 mW whereas the length of the resonators is between 0.1 and over 1 m. The coherence length can be estimated from the width of the gain curve of around 1.5 MHz. According to Eq. (7.9) a length of 50 cm results in $N = 10$ modes inside the resonator.

This results in a coherence length of about 10 cm, Eq. (7.10). Experimental values are a little bit larger.

The beam diameter of He–Ne laser is about 0.7 mm; for commercial lasers mostly the TEM$_{00}$ mode occurs so that the beam can be described by a Gauss profile. The divergence is roughly $\Theta = 0.6$ mrad, which means the beam diameter increases by 1.2 mm for 1 m distance. For holography it is recommended to use lasers with polarized radiation (Section 2.1).

7.3.2
Ion Laser

Ion lasers using the inert gases argon and krypton deliver much higher output powers whereas in holography often lasers with several watts are used (Table 7.1). The gain profile is highly broadened by the Doppler effect due to the high current densities leading to line widths of 6 GHz and more. Compared to He–Ne lasers the coherence lengths of common ion lasers are small, which is why these types are only usable for holography with an etalon inside the resonator.

Figure 7.6 shows the setup of a typical argon-ion laser. It emits in the blue and green spectral region (Table 7.1). A single line can be selected by selective mirrors or by a prism inside the resonator. Additionally a mode aperture and an etalon are necessary to switch to transversal and longitudinal monomode operation.

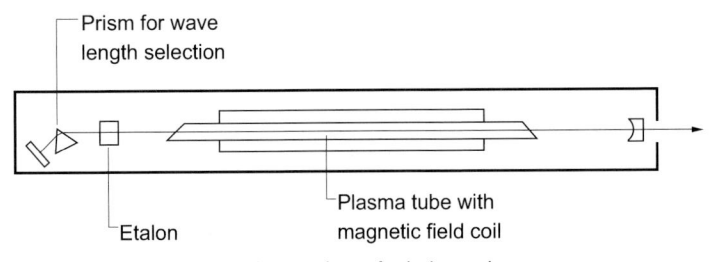

Fig. 7.6 Setup of a typical argon laser for holography.

The etalon is slightly tilted to prevent that it acts as an end mirror. The adjustable tilt also has the advantage that the resonance frequency of the etalon coincides with the line center and thus optimizing the output power.

A typical argon laser for holography has an output power of about 2 W in the strongest line (green, 0.514 µm). This is reduced by the etalon to about 1 W. The laser needs the electrical power of about 13 kW and uses a three-phase current (32 A) and a water connection with a relative high flow-through (9 l/min).

Krypton lasers are similar to argon lasers. For the same electrical power the laser power is reduced to a third. But for holography with AgBr layers the krypton laser is superior to the argon laser because the sensitivity of the layer is 10 times higher for red (0.647 µm) than for green. For working with photoresist the krypton laser is unsuitable.

7.3.3
He–Cd Laser

Although it is an ion laser this laser type is more like a He–Ne laser despite the fact that its wavelength is in the blue region at 0.442 µm. Its power is limited to about 100 mW. For the He–Cd laser the use of photoresist layers is favorable because they exhibit a high sensitivity in the blue spectral region.

7.4
Solid-State Lasers for Holography

Solid-state lasers emit short pulses in the ns region [15]. Therefore the exposure time for the recording of the hologram is so short that problems with stability practically do not occur. Most often used is the ruby laser which generates a high pulse energy of up to 10 J. The red radiation (0.694 µm) lies in the sensitivity maximum of red sensitive holographic AgBr films. Additionally, Nd:YAG lasers are used for holography because they have a higher efficiency and higher pulse frequencies. But for holography the infrared radiation of the laser (1.06 µm) needs to be frequency-doubled with a crystal into the green region (0.53 µm).

7.4.1
Ruby Laser

The setup of a ruby laser for holographic applications is shown in Fig. 7.7. The active medium consists of a ruby crystal (Al_2O_3 doped with 0.05% Cr_2O_3 or $Al_2O_3:Cr^{3+}$) with a diameter of a few mm. The crystal is optically pumped using light from a xenon flashlight. The crystal and the linear flashlight are enclosed in a water cooled elliptical pump chamber. For the use in holography the laser is equipped with a variable attenuator (q-switch). This setup represents a fast optical shutter. At the beginning of the pump pulse with a duration of 0.1 ms the resonator is closed so that no laser radiation is created. Once the population inversion reaches its maximum the q-switch is opened. The high inversion is released within a few ns emitting a giant pulse with a maximum power in the MW region. In the extreme case the pulse duration τ resembles the time for one light cycle inside the resonator of length L:

$$\tau \approx \frac{2L}{c}. \tag{7.13}$$

For a resonator of length 0.6 m the pulse duration is estimated to $\tau = 4$ ns whilst it is a bit longer in reality. The dependence of pulse energy W and pulse power P is (for rectangular pulses) given by

$$P = \frac{W}{\tau}. \tag{7.14}$$

For a pulse energy of 1 J and a pulse duration of 4 ns the calculated pulse power is $P = 250$ MW.

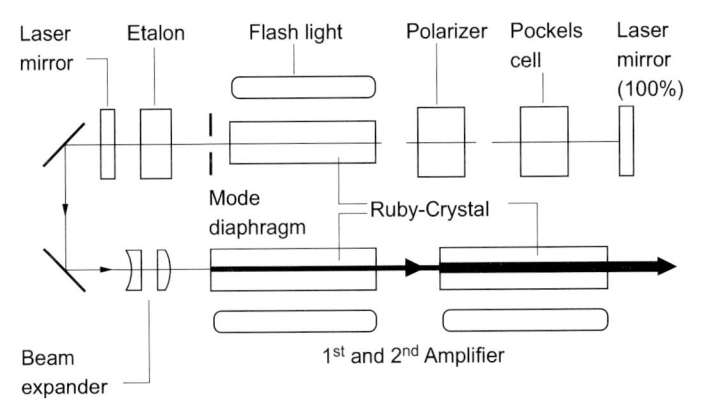

Fig. 7.7 Lay-out of a holographic Ruby laser

The q-switch for solid-state lasers consists of a polarizer and a pockels cell inside the resonator. A glass plate positioned at the Brewster angle can be used as a polarizer. The pockels cell consists of an electro-optical crystal, e.g., KDP, for which the crystal axis has an angle of 45° with respect to the polarization of the laser beam. Right before igniting the flashlight a voltage of a few kV is applied to the crystal. Thus it becomes birefringent and generates circular polarized light. After reflection at the end mirror the light again passes the crystal and is again linearly polarized but rotated by 90°. The light suffers losses in the resonator. Thus the resonator is closed and the laser does not start to oscillate. At the end of the pump pulse the voltage is switched off so that the light is able to pass. The laser oscillates and emits a giant pulse.

Solid-state lasers usually emit a large number of transversal modes so that spatial coherence is small. Lots of irregular structures are observable in the beam profile due to the superposition of the modes. By adding a mode aperture higher modes can be suppressed leaving only the TEM_{00} mode.

The fluorescence line of a ruby laser is broadened up to 330 MHz due to (homogeneous) lattice vibrations. According to Eq. (7.9) therefore several hundred longitudinal modes are formed and the coherence length is smaller than 1 mm. Therefore, it is necessary to use an etalon (several mm thick) for holographic applications making it possible to create coherence lengths up to the meter region.

With increasing pump power the gain curve grows above the laser threshold and the number of modes increases. Thus the pump power must not be too high for monomode operation and the pulse energy has to be limited to about 50 mJ. The energy can be increased by using one or two amplifiers, which consist of ruby rods with diameters up to 20 mm and lengths of 200 mm. The diameter of the beam is expanded by a telescope. Systems with one amplifier are able to deliver an energy up to 1 J whilst with two amplifiers up to 10 J are reached.

7.4.2
Nd:YAG Laser

The setup of a Nd:YAG laser is much like that of the ruby laser. The laser crystal consists of $Y_3Al_5O_{12}$ doped with neodymium (Nd) in a weight proportion of 0.7%. The abbreviation "YAG" stands for yttrium aluminum garnet. The laser oscillator generates pulses of small energy but large coherence length which are intensified by laser amplifiers. Suitable oscillators are especially systems that can be pumped with laser diodes. In this case a homogeneous illumination of the crystal is achieved and the heating is highly reduced which improves the beam quality.

The radiation of neodymium lasers is in the infrared region at 1.06 µm. To be used in holography a conversion of the radiation to visible light is needed. Therefore a crystal for frequency doubling (e.g., KD*P) is mounted at the exit of the whole laser system. By exact adjustment of the crystal 50% and more of the radiation can be converted to green light at 0.53 µm. The green light is separated by a selective mirror from the infrared light.

7.5
Lenses and Spatial Filters

In holography the laser beam with a diameter in the mm region is expanded by a system of lenses to 10 cm and more. When calculating the optical path the wave property of the light has to be taken into account.

7.5.1
Gaussian Beam

Lasers for holography emit in the TEM_{00} or Gauss mode. Such a beam has a minimal diameter w_0 from which the beam propagates with a certain divergence (Fig. 7.8). A lens transforms a Gauss beam with w_0 into another Gauss beam with w'_0. The following equations apply [14, 16]:

$$a' = -f + \frac{f^2(f-a)}{(f-a)^2 + z_R^2} \qquad (7.15)$$

and

$$\frac{w'_0}{w_0} = \frac{f}{\sqrt{(a-f)^2 + z_R^2}} \quad \text{with} \quad z_R = \frac{\pi w_0^2}{\lambda}. \qquad (7.16)$$

The parameters are explained in Fig. 7.8. The position of the beam waist before and after the lens is defined by a and a'. (The quite unusual sign convention (a positive, a' negative) is because the wavefronts have different curvatures before and after the lens.) The Gauss beam has a divergence angle of $\Theta = \lambda/\pi \cdot w_0$ which is defined in Fig. 7.9. The focal depth is $b = 2z_R$.

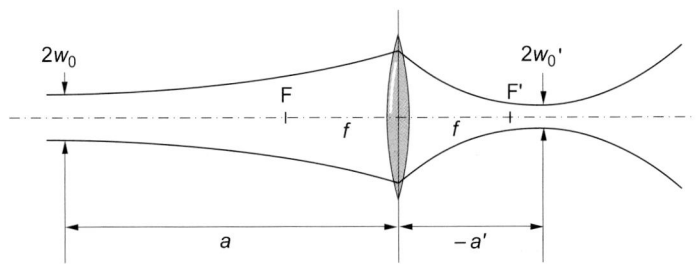

Fig. 7.8 Propagation of a Gauss beam (TEM$_{00}$ mode) through a lens (sign convention: a positive, a' negative).

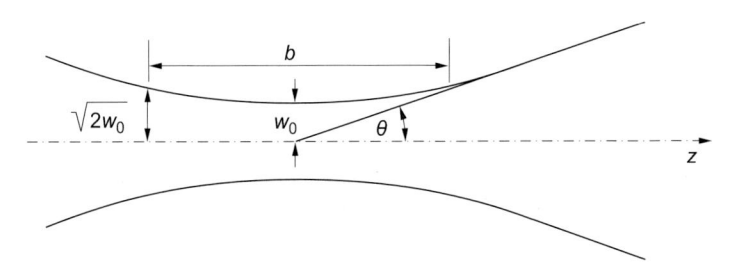

Fig. 7.9 Divergence of a laser beam (TEM$_{00}$ mode).

7.5.2
Focusing

In many applications a focusing of the laser beam is needed. For that z_R (and w_0) needs to be as large as possible according to Eq. (7.16). In this case the focal radius is given by

$$w_0' \approx \frac{\lambda \cdot f}{\pi \cdot w_0},\tag{7.17}$$

where w_0 as an approximation that can be set equal to the beam radius at the lens.

7.5.3
Geometrical Optics

Often the beam propagation can be calculated approximately by using geometrical optics. For $z_R \ll a - f$ the equations for the Gauss beam transform to the usual imaging equations:

$$\frac{1}{f} = \frac{1}{s_i} + \frac{1}{s_o}\tag{7.18}$$

and

$$\frac{y_i}{y_o} = \frac{s_i}{s_o}.\tag{7.19}$$

Here a lens is characterized by two main planes H an H' which coincide for thin lenses, see Fig. 7.10. The following sign convention has to by obeyed which differs from that of the Gauss beam: for object to the left of the lens the object distance s_o is positive, and for images to the right the image distance s_i is positive.

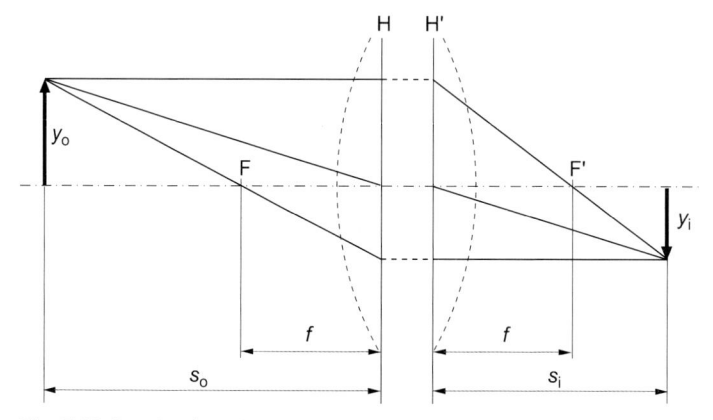

Fig. 7.10 Imaging by a lens according to the laws of geometrical optics (s_o = object distance, s_i = image distance, y_o = object size, and y_i = image size).

As an example the focusing of a He–Ne laser beam with a radius (in the middle of the laser) of $w_0 = 0.3$ mm through a microscope objective (40×,

$f = 4$ mm) at a distance $s_o = 1$ m shall be calculated. According to the theory of the Gauss beam this yields $z_R \approx 500$ mm, $a' \approx -f \approx -4$ mm, and $w'_0 = 1.2$ µm. The result of the geometrical optics is nearly the same: $s_i \approx f$ and $y_i = w'_0$. But not in all cases the result is a good approximation.

7.5.4
Spatial Filters

Laser light in practice is superposed with irregular interference structures caused by diffraction at dust and scratches in the optical system. In holography it is often necessary to "clean" the beam in order to achieve a homogeneous intensity distribution according to the TEM$_{00}$ mode. This is done by a spatial frequency filter (Fig. 7.11). In such filters the laser beam is focused through a lens, often a microscope objective. In the created beam waist in the vicinity of the focal plane an adjustable pinhole is positioned with an opening larger than the spot diameter. Thus the TEM$_{00}$ radiation can pass the pinhole unhampered. Small dust particles create diffraction images of large diffraction angle, i.e., of large spatial frequencies. In consequence, the diffracted waves do not pass the aperture anymore. Thus undesirable waves are spatially separated from the Gauss beam and are masked out.

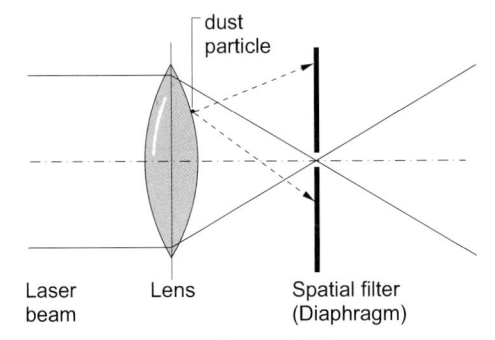

Fig. 7.11 Principle of a spatial filter.

The diameter d of the aperture has to be larger than the diameter of the beam waist in the vicinity of the focal plane (Eq. (7.17)):

$$d > \frac{2f \cdot \lambda}{\pi \cdot w}, \tag{7.20}$$

where w denotes the beam radius of the incoming radiation. On the other hand, d needs to be small so that side maxima cannot pass the aperture. This is the case when d is smaller than the diameter of the so-called first Fresnel zone:

$$d < 2\sqrt{f\lambda}. \tag{7.21}$$

In practice, d will be chosen as large as possible within these limits in order to reduce difficulties in adjustment. Typical values for a spatial filter used in holography are around $d \approx 40$ µm. Microscope objectives with $f = 4$ mm are used (40×).

7.5.5
Beam Expansion

Behind the spatial filter the radiation propagates divergent and can be directly used as a reference or illumination wave. In many cases though a parallel beam is desired. For this purpose a convex lens with a large diameter is used behind the objective. If the two focal planes coincide this setup represents a Kepler telescope and a parallel beam is created (Fig. 7.12). According to the laws of geometrical optics the beam diameter is increased from d to D:

$$\frac{D}{d} = \frac{f_2}{f_1}. \tag{7.22}$$

Using the beam expansion after Kepler with lasers of high power an electrical flash over in air can occur in the focus of the objective. In this case it is advised to use a beam expansion after Galilei where a concave lens replaces the objective (Fig. 7.12).

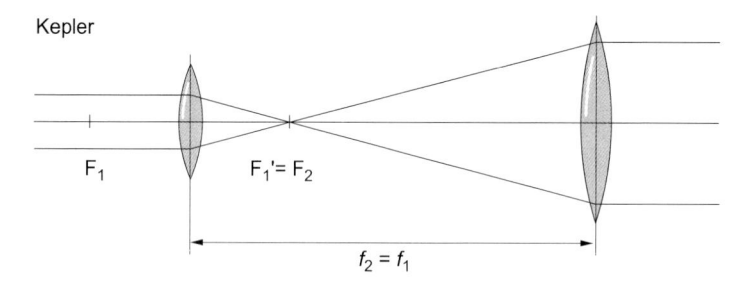

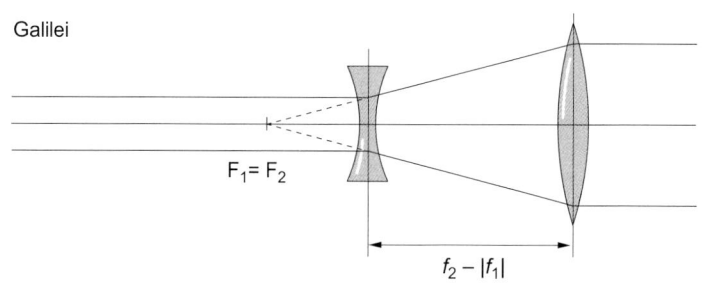

Fig. 7.12 Beam expansion using a Kepler telescope and a Galilei telescope.

7.6
Polarizers and Beam Splitters

Light represents a transversal electromagnetic wave in which the electric and the magnetic field strengths oscillate orthogonal to the propagation direction. The magnetic field strength is of minor importance for the interaction with matter; for holography therefore only the electric field strength is relevant.

7.6.1
Polarization

The term "polarization" describes the orientation of the oscillation plane of the light. Polarization is distinguished between the linear and elliptical one; circular polarization is considered as a special case of the elliptical one. In unpolarized radiation all oscillation directions occur statistically. Common light sources emit unpolarized light whilst lasers normally emit linear polarized or unpolarized light. For holographic applications usually linear polarized light is used where the direction of polarization should be parallel to the plane of incidence.

Linear polarized light can be generated from elliptical or unpolarized radiation by using dichroitic filters, birefringence, or reflection. Birefringent $\lambda/4$-plates transform linear polarized light into circular polarized light. A rotation of the plane of linear polarized light can be done with a $\lambda/2$-plate.

7.6.2
Dichroitic Filters

For generation of linear polarized light often dichroitic polarization filters are used. "Dichroism" means that light of one polarization direction is selectively absorbed. The disadvantage of these filters is that also for the desired polarization direction losses are quite large.

7.6.3
Polarization by Reflection

Light that is reflected by a surface is partly polarized. On metallic surfaces usually an elliptical polarization occurs; the dependences are quite complicated. Of relevance for holography is also the reflection on glass surfaces, e.g., the glass plates which carry the holographic layer. To describe the reflection the light is divided into two components, parallel and perpendicular to the incidence plane. In Fig. 7.13 it is observable that light which is polarized perpendicular to the incidence plane is more strongly reflected than parallel polarized light. Thus to reduce reflection the radiation needs to be polarized parallel to the incidence plane. For the Brewster angle ($\tan\Theta_p = 1/n \approx 56°$)

this reflection vanishes completely. It is favorable to set up glass plates at this angle in the optical path. This prevents interference of the waves reflected at these two surfaces if the polarization direction is right. Glass plates at the Brewster angle are also used in resonators, e.g., in solid-state lasers, in order to polarize the radiation.

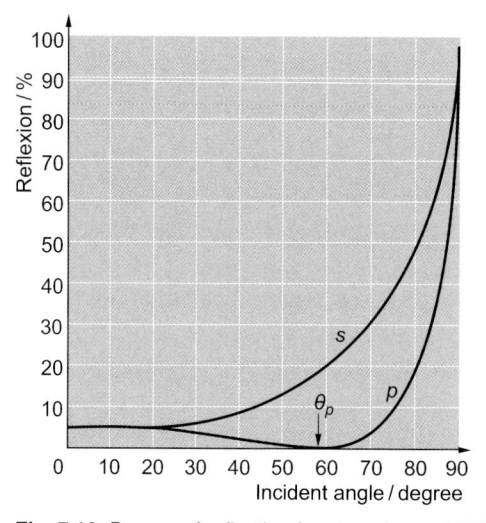

Fig. 7.13 Degree of reflection for glass ($n = 1.52$) for different polarization directions (s = polarization perpendicular to the incidence plane, p = parallel).

7.6.4
Polarization Prisms

Birefringent prisms for polarization are used in different versions. The often mentioned Nicol prism is sparely used nowadays. The Glan–Taylor and Glan–Thompson prisms convert an unpolarized beam into a polarized one without changing its direction (Figs. 7.14 and 7.15). The other polarization is totally reflected and the beam is mirrored out of the optical path. When calcite is used prisms can be used for wavelength between 0.3 and 2.3 µm. Polarization prisms are allowed for higher powers if the two prisms are separated by an air gap. Conventional prisms are joined with a putty which can be destroyed by high laser power. The Wollaston prism separates an unpolarized beam into two beams which exit the prism in different directions. The setup for a variable beam splitter using a Glan–Taylor prism is shown in Fig. 7.15.

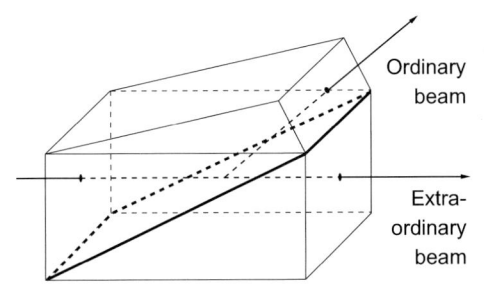

Fig. 7.14 Polarization prism after Glan–Thompson.

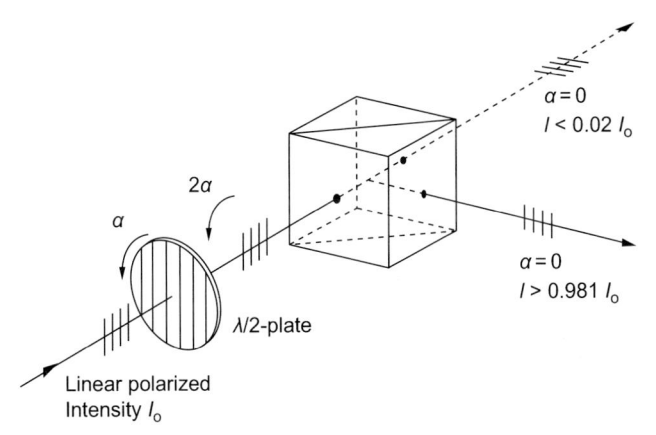

Fig. 7.15 Setup of a lossless variable beam splitter using a Glan–Taylor prism and a $\lambda/2$-plate. By a second $\lambda/2$-plate not shown here the polarization of both beams can be rotated in the same direction.

7.6.5
Thin Film Polarizers

Thin film polarizers are similar to dielectric mirrors made of several layers. They are passed by the light at an angle and the light is polarized that way.

7.6.6
$\lambda/4$- and $\lambda/2$-plates

The generation of circular polarized light is done with $\lambda/4$-plates made from mica or quartz. The plates exhibit a crystal axis in the film plane which should be marked. The direction of the linear polarization of the incident light has to have an angle of $45°$ to this preferred axis. The indices of refraction in the direction of the crystal axis and perpendicular to it are different so that two waves with different speeds exist. Thus a phase difference between the two is introduced. The thickness of the plates is chosen in a way that the effective path difference is $\lambda/4$ and circular polarized light is created (Fig. 7.16). Using a polarizer with a $\lambda/4$-plate an optical isolator can be constructed which passes light in one direction and reduces reflected light.

The rotation of linear polarization is done with $\lambda/2$-plates. By rotation of the plate the angle of polarization can be varied continuously (Fig. 7.17). If the polarization of a laser is in the wrong direction it can be turned by such a plate so that it coincides with the hologram plane.

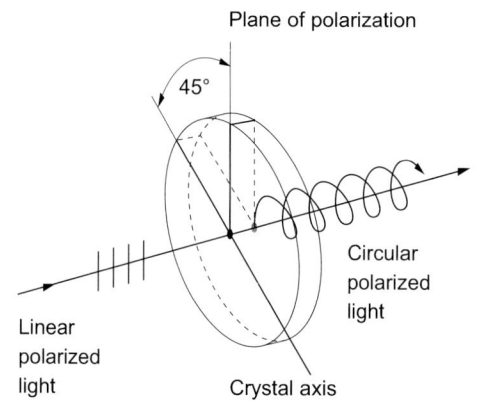

Fig. 7.16 Working principle of a $\lambda/4$-plate. Incident linear polarized light is transformed into circular polarized light.

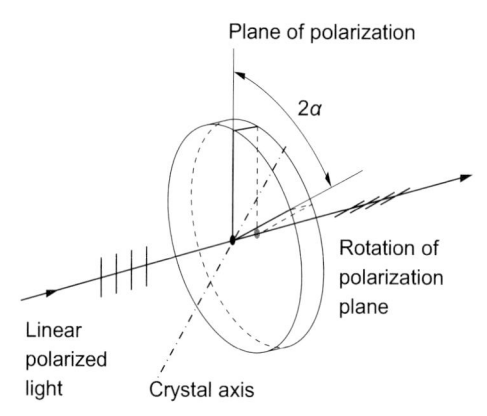

Fig. 7.17 Working principle of a $\lambda/2$-plate. The polarization plane can be rotated by every angle 2α where α is the angle of rotation.

7.6.7
Beam Splitter

Figure 7.15 shows a variable lossless beam splitter as it is used in holography. Linear polarized light falls onto a polarization prism which delivers two perpendicular beams of different polarization. By rotating the $\lambda/2$-plate from 0°

to 45° the polarization of the incoming beam is turned from 0° to 90° and the intensity can be split between the two beams as needed. Another $\lambda/2$-plate (not shown) can rotate the polarization of the two exiting beams into the same direction.

Another type of beam splitter uses a glass plate which is put into the optical path at 45° or a different angle. The front side consists of a dielectric half transparent mirror with reflection coefficient $0 < R < 1$. It has to be noted that the reflection coefficient is specified for 45°. The backside is mostly broadband nonreflecting; the absorption is around 0.5%.

Pellicle beam splitters are thin, stretched foils of nitrocellulose that can be vaporized with dielectric layers. Their advantage is that a negligible geometric-optical beam offset for the passing beam occurs. However, these beam splitters are mechanically delicate.

As a beam splitter also dielectric or metallic layers can be used that are put diagonally into a joined glass cube. Depending on the type the polarization components can be reflected with the same or with different intensities.

7.6.8
Metal Mirrors

The physics of light-reflection at metal surfaces is complicated. For an angle incoming linear polarized light and the reflected light can be elliptical polarized. The nonreflected part is absorbed in the mirror. High degrees of reflection of 99% occur for infrared light whilst for visible light 91% are reached for Al and 98% for Au. Metal mirrors are often vaporized onto glass carriers and sometimes covered with protective coatings (e.g., MgF_2 and SiO_2).

7.6.9
Dielectric Multilayer Mirrors

By adding thin layers onto optical surfaces their reflection properties can be strongly changed. Interferences at these layers lead to reflecting or nonreflecting properties in a narrow wavelength region.

7.6.10
Nonreflective Coating

According to Fig. 7.13 about 4% of the light is reflected at the glas–air boundary layer for perpendicular incidence. By vaporizing the glass surface with a dielectric layer of the optical thickness $nd = \lambda/4$ this reflection can be reduced or even prevented for one specific wavelength. To do so the refraction index n of the layer has to fall between that of air and that of the glass. The decrease of reflection is due to the interference of the two waves reflected at the front side and the back side of the $\lambda/4$-layer. Lenses are often coated routinely with $\lambda/4$-layers made of MgF_2.

7.6.11
Laser Mirrors

Low loss mirrors with a high degree of reflection can be created by using a coating of several $\lambda/4$-layers. Already a single dielectric coating on a substrate can increase the reflection drastically. Contrary to the conditions for a reflection decrease the layer has to have a refraction index n that is larger than that of the glass. Thus the phase jump at the boundary layer coating glass is omitted whilst it occurs with a value of π at the layer air-glass. The overall path difference of the two reflected waves is now λ so that they superpose constructively. A highly diffractive layer of ZnS ($n = 2.3$), e.g., increases the degree of reflection of glass ($n = 1.5$) from 4% to 31%. Higher degrees of reflection of over 99% are achieved with multilayer mirrors which can be almost lossless; they consist of interchanging high and low diffractive transparent $\lambda/4$-layers.

7.7
Vibration Isolation

The optical setup has to be mechanically so stable during the holographic recording that the movement of the interference fringes is small compared to the light wavelength. Already vibration amplitudes of $\lambda/20 \approx 30$ nm noticeably decrease the contrast of the interference fringes and the brightness of the holograms. When applying pulse lasers in q-switch, e.g., ruby lasers, this is not a problem since the exposure time is given by the pulse length of a few ns ($= 10^{-9}$ s). The use of continuous gas lasers increases the exposure time to several 0.1 s up to minutes. In this case mechanical vibrations of the building can cause blurring of the interference fringes in the hologram plane. Therefore it is necessary to use vibration isolating tables when working with continuous lasers. The isolation of the vibrations has to be dimensioned in a way that the amplitudes in the holographic setup are maximal in the nm region. Causes of vibration can be running machines, air conditioners, traffic, or moving persons.

A vibration isolating table usually consists of two parts which both play an important role for the isolation:

Isolators: The tabletop is placed onto vibration isolators which can be constructed in several ways. They decrease the transfer of vibrations to the optical setup.

Table top: The construction of the table top can be chosen such that further absorption of vibrations is achieved. This holds especially for metal tops with honeycomb structures. More simple tops are made of natural or artificial stone. The isolators have to be adjusted to the mass of the table top.

7.7.1
Isolators

For a holographic table the table top is placed vibratory onto isolators. As isolators steel springs or pneumatic elements are used. The latter use compressed air or they are closed circuit rolling bellows. For simpler setups also motorcycle or car tire tubes can be used. The system of isolators and table top can be modeled by a spring with an added mass and a damper. Such oscillatory systems can be described by the equations for forced oscillations [17]:

$$T = \frac{x}{x_0} = \frac{1}{\sqrt{(1-(\frac{f}{f_0})^2)^2 + (\frac{d \cdot f}{f_0})^2}} \qquad (7.23)$$

with

$$d = \frac{\beta}{\pi f_0}.$$

Here T defines the degree of transmission which describes the ratio of the amplitude of the table top oscillation x to that of the driving oscillation x_0. The system is driven with oscillations of frequency f and has an eigenfrequency of f_0 (for the undamped case). The parameter β which describes the oscillation decrease after switching of the driving force is called the "decay constant." Instead of β often the dimensionless constant $d = \beta/\pi f_0$ is used, the so-called dissipation factor.

Figure 7.18 shows the transmission of an oscillatory system according to Eq. (7.23). A holographic table should exhibit $d = 0.5$ to 2. For a small damping the amplitude increases in the resonance. The value $d = 2$ ($\beta = 2\pi f_0$) defines the case of aperiodic damping which ensures a fast return of the table top into its resting position after an external disturbance.

In Fig. 7.18 it is observed that the resonance frequency is approximately the same as the eigenfrequency f_0. Above the resonance frequency $f \gg f_0$ the transmission decreases with $1/f^2$, i.e., a doubling of the driving frequency decreases the amplitude by a factor of 4.

A holographic table should have an eigenfrequency of $f_0 = 1$ to 2 Hz and be damped aperiodic. For this purpose the isolators have to be adjusted to the mass of the table top. A vibration isolator or a spring is characterized by the so-called spring constant D, which is defined by the relation force = deflection $\times D$. The resonance frequency of an oscillator is given by the spring constant

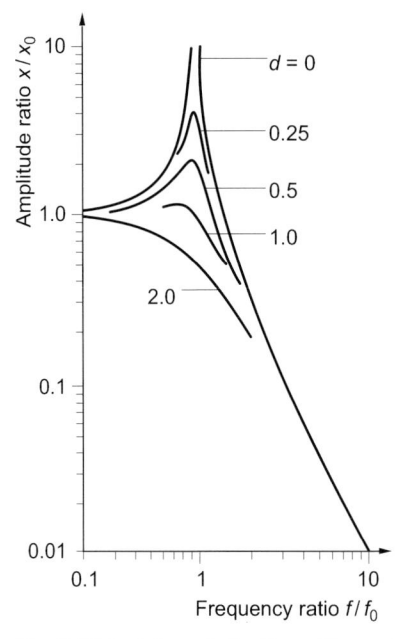

Fig. 7.18 Isolation of vibrations. Transmission of an oscillatory system for different factors d according to Eq. (7.23). The curve with $d = 0.5$ approximately represents the behavior of a holographic table.

D and the mass m with which it is loaded:

$$f_0 = \frac{1}{2\pi} \cdot \sqrt{\frac{D}{m}}. \tag{7.24}$$

For isolators using steel springs D is constant; thus the eigenfrequency decreases with increasing load. More favorable are the more often used pneumatic isolators. They consist of a piston which can move inside a cylinder under high pressure. The spring constant in this case can be calculated as follows:

$$D = \chi m g \sqrt{\frac{A}{V}}. \tag{7.25}$$

In the above equation $\chi \approx 2$ denotes the adiabatic coefficient, $g = 9.81$ m/s² the gravity constant, V the gas volume of the cylinder and A the piston area. It is obvious that the spring constant is proportional to the mass m. Inserting Eq. (7.25) into Eq. (7.24) one gets an eigenfrequency independent of the mass. Precondition for doing so is that the gas volume V is kept constant. This can be achieved by an automatic or manual pressure regulation. Especially suitable for holography are isolators that are connected to a compressed air system and where the volume is kept constant by automatically adjusting the piston. When using motorcycle or car tire tubes the volume can be adjusted by the pressure and therefore f_0. Another alternative are rolling bellow isolators which are inflated by an air pump or by compressed air depending on the load.

The damping of the oscillation is achieved by an additional gas volume that is connected to the main volume by a defined opening. When using tire tubes the damping is done by the rubber mantle.

7.7.2
Table Tops

Vibrations passing the isolators should be efficiently damped inside the table top. Simplest of this is realized by using tops with a mass as high as possible. However, this has practical limits. Another solution is the use of steel honeycomb structures which are sandwiched between 5 mm thick bottom and cover plates.

To characterize the quality of the table tops, i.e., as a measure for the stiffness and internal damping, the term "compliance" is introduced. It describes the reciprocal of the spring constant of the plate which is defined by the oscillation amplitude at one point of the table caused by a driving force (compliance $C =$ amplitude$x$/force$f$). Since the oscillation amplitude is measured at a corner of the table one talks of "corner compliance." For a simple system with only one resonance frequency the compliance C can be described mathematically by transformation of Eq. (7.23). In practice though several oscillation modes with different resonance frequencies occur, so that this transformation is not of use.

In Fig. 7.19 the compliance of a table top with honeycomb structure is shown as a function of the driving frequency. The strong decrease down to about 100 cycles/s is proportional to $1/f^2$ and is due to the mass inertia. The internal damping of the system is represented by the decrease of the curve at high frequencies and the flattening of the present resonances. The resonances should be located far beyond 100 cycles/s; otherwise high amplitudes would occur. In the example shown the first resonance occurs at 150 cycles/s, the maximum compliance is 4^{-5} mm/N.

7.7.3
Vibration Isolated Table

The behavior of a holographic table is governed by the isolators and the table top. The mass of the table top, the isolators and, in the case of a pneumatic suspension, the pressure have to be adjusted in a way that the resonance falls between 1 and 2 Hz. Additionally every disturbance of the system should come back to the resting position aperiodically. In Fig. 7.20 the transmission

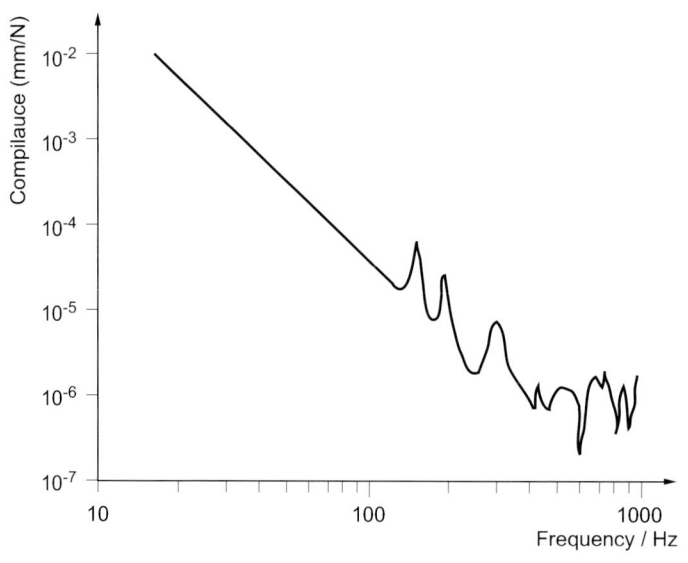

Fig. 7.19 Corner compliance of a holographic table top with honeycomb structure (2.4×1.2×0.2 m). Clearly visible are the resonance frequencies which should be damped as good as possible.

of a holographic table with the honeycomb structure is shown. Here it is distinguished between vertical and horizontal vibrations. In the vertical direction the transmission is a bit less since the pistons move in that direction; the resonance frequency in this case is lower (a in Fig. 7.20). The damping in the horizontal direction is smaller leading to a slower decrease of the compliance (b in Fig. 7.20). At high frequencies the resonance frequencies of the table top can be observed.

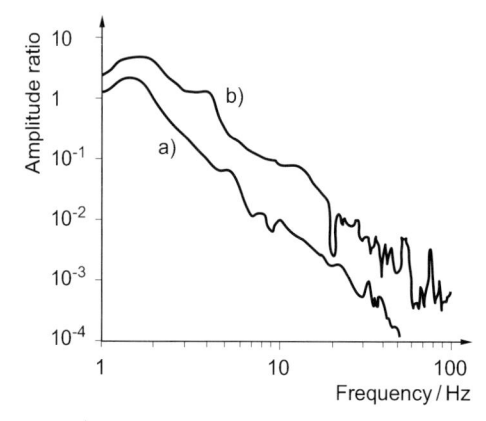

Fig. 7.20 Transmission of a holographic table with vibration isolators: (a) vibrations in the vertical direction and (b) vibrations in the horizontal direction.

7.8
Optical Fibers and Diode Lasers

In the past few years significant simplifications in the technique of holography have been achieved. The application of optical fibers simplifies the beam guidance; the development of diode lasers with long coherence lengths reduces the technical expense and the price of light sources, mainly for the holographic metrology.

7.8.1
Monomode Fibers

Optical fibers made of quartz glass are used in several applications to transport laser light. For holography so-called monomode fibers are suitable with a core diameter in the µm region which had been developed mainly for applications in information and measurement technologies. Normal and polarization maintaining fibers are used. To couple in the light from the laser it is focused by an objective on the end plane of the fiber. For this task prefabricated precision-mechanic complete systems are available including fibers of different lengths with added plug systems.

Figure 7.21 shows a holographic setup in which the reference and illumination waves are guided by monomode fibers. The advantage lies in the flexibility of the setup which can be altered quickly. The wave is divided by a beam splitter and then guided into the fiber by two adjustment systems. The lengths of the fibers are chosen in a way that the optical paths to the hologram are nearly the same. The coherence is maintained in a monomode fiber and a divergent beam with a Gauss-like profile exits at the end plane. The

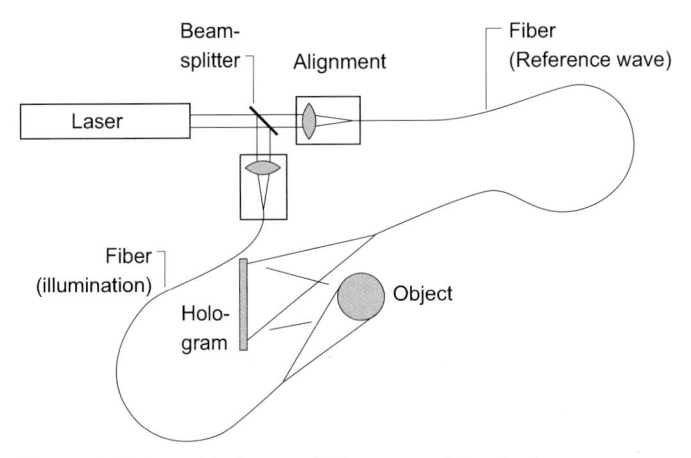

Fig. 7.21 Holographical setup. Reference and illumination wave are guided by fibers.

opening angle depends on the numerical aperture of the fiber, typically being around 10°. For better illumination of the object a third fiber with an additional beam splitter can be used. The demands concerning stability are the same as with other common setups, the advantage being the easier handling. It has to be taken care of that the end planes of the fibers are kept clean and are not touched. Until now their use is limited to laser powers of up to 1 W; pulse lasers cause damages to the fiber surface during coupling.

7.8.2
Diode Lasers

Diode lasers with small excitation can also oscillate in the longitudinal monomode. This leads to a large coherence length and to an application in holography. Until now especially GaAs lasers with several mW in the near infrared were used. The radiation is not visible and the holograms have been taken with holographic polaroid cameras and made visible with CCD cameras. The applications are in the holographic metrology and interferometry.

One disadvantage of diode lasers is the large divergence of the emitted coherent radiation. Due to diffraction the radiation is emitted in a cone with opening angles between 5° and 40° [14, 18].

Problems

Problem 7.1 Show that the intensity in the interference pattern of Fig. 7.1 (assumed complete spatial coherent light) oscillate between $I_{max} = 4I_0$ and $I_{min} = 0$.

Problem 7.2 Calculate the coherence length l_c of a He–Ne laser ($\Delta f = 1$ GHz) and white light ($\Delta \lambda = 250$ nm, mean wavelength $\lambda = 500$ nm).

Problem 7.3 After a distance of $l = 20$ m the 632.8 nm He–Ne laser beam has a diameter of 8 cm. Calculate the divergence angle and the beam diameter w_0.

Problem 7.4 A He–Ne laser of 1 m length has seven longitudinal modes. Calculate the coherence length and the bandwidth of the coherent radiation.

Problem 7.5 An Ar laser of $L = 2$ m length and 1 W optical power has a coherence length of 5 cm at $\lambda = 488$ nm. Calculate the total bandwidth of the 488 nm radiation. What thickness of etalon (refractive index 1.5) is needed to reduce the number of longitudinal modes to one (monomode). Calculate the finesse F, the reflectivity of the etalon surfaces and the coherence length of the laser with etalon included.

Problem 7.6 A He–Ne laser has a divergence of 1 mrad. At a distance from the laser exit of $d = 1$ m the laser beam is focused with a microscope objective

($40\times$, $f = 4$ mm). Calculate the radius of the focus w'_0. Try to solve the problem using geometrical optics.

Problem 7.7 A laser beam of a diameter of $d = 1$ mm shall be expanded to give a plane wave of $D = 15$ cm diameter. Using a microscope objective ($40\times$, $f_1 = 4$ mm), what is the focal length f_2 of the lens needed.

8
Basic Experiments in the Holographic Laboratory

This chapter describes a choice of introductory experiments which can be performed in the holographic laboratory. The following topics are covered: experiments in geometric and wave optics, problems of mechanical stability, working with film material and simple holographic experiments.

8.1
Polarization and Brewster Angle

8.1.1
Experiment 1: Analyzer and Polarizer

An unpolarized light ray (light bulb or laser) is created on an optical bench (Fig. 8.1a) and is linear polarized by a polarizer, i.e., a polarizing foil or prism. The proof of the polarization is done by rotating an analyzer which is similar in construction as the polarizer. The intensity of the light can be varied between bright and dark. The analyzer transfers only that part of the light amplitude (electrical field strength) which is parallel to its preferred direction. Since the intensity I is proportional to the square of the field strength E the result for the rotation is

$$I = I_0 \cos^2 \alpha, \tag{8.1}$$

where I_0 is the maximum intensity and α is the rotation angle with respect to the parallel position of the polarizer and analyzer. If a laser with polarized radiation is used in the experiment the polarizer is unnecessary.

8.1.2
Experiment 2: Rotation of the Polarization Plane

By adding a $\lambda/2$-plate in a polarized beam the polarization plane can be changed. A rotation of this element by an angle α changes the polarization direction by 2α which can be proved with the analyzer (Figs. 8.1b and 7.17). For white light each color is rotated by a different angle so that light of different colors is created behind the analyzer. Some lasers exhibit a polarization

Holography: A Practical Approach. Gerhard K. Ackermann and Jürgen Eichler
Copyright © 2007 WILEY-VCH Verlag GmbH & Co. KGaA, Weinheim
ISBN: 978-3-527-40663-0

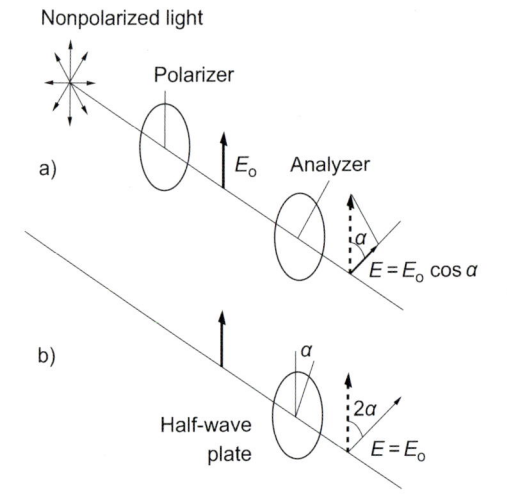

Fig. 8.1 Experiments on linear polarization: (a) effect of analyzer and polarizer (for linear polarized light the polarizer can be omitted) and (b) rotation of the polarization direction using a $\lambda/2$-plate; the rotation angle can be measured with an analyzer.

parallel to the plane of the holographic table. By using a $\lambda/2$-plate (at 45°) the polarization can be rotated vertical; thus it is parallel for two divided beams so that interference fringes exhibit maximum visibility (Section 2.1).

8.1.3
Experiment 3: Brewster Angle

Light reflected by a glass plate is partially polarized since according to Fig. 7.15 the reflection coefficient is a function of the polarization of the incident light. At the Brewster angle Θ_p the reflected light is completely polarized. This angle is given by

$$\tan \Theta_p = n. \tag{8.2}$$

For common glass the index of refraction is about $n = 1.5$; accordingly the Brewster angle is $\Theta_p = 56°$. In experiment 3 a glass plate is setup such that the polarization plate is perpendicular to the plate (Fig. 8.2a). After that the plate is rotated according to Fig. 8.2a and the intensity of the reflected light which is given by the degree of reflection of component p in Fig. 7.13 is observed. At the Brewster angle the reflected beam vanishes.

If the polarization is chosen such that the direction is parallel to the plate or if unpolarized light is used the reflected beam is completely polarized when selecting the Brewster angle (Fig. 8.2b). This can be proved with a polarizing filter.

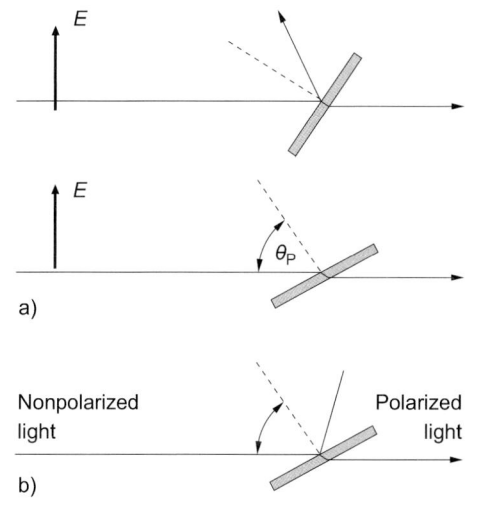

Fig. 8.2 Experiments on the Brewster angle Θ_p: (a) for the Brewster angle no reflection occurs for the polarization direction (in the direction of the incidence plane) shown and (b) For the Brewster angle the reflected light is completely polarized (perpendicular to the incidence plane).

In holography it is favorable to setup the hologram films and glass substrates at the Brewster angle because reflections and interferences caused by those are omitted (see experiment 8).

8.1.4
Experiment 4: Variable Beam Splitter

With the help of a polarizing prism and a $\lambda/2$-plate a variable beam splitter according to Fig. 7.15 can be setup. The splitting ratio can be adjusted by rotating the $\lambda/2$-plate. If all surfaces are nonreflecting this beam splitter works lossless. By using a second $\lambda/2$-plate not shown in Fig. 7.15 the polarization of the divided beams can be rotated in the same direction.

8.2
Experiments with Lenses

8.2.1
Experiment 5: Measuring Focal Length

For the experiments with lenses a long optical bench is setup. An illuminated metal sheet with a letter made of holes serves as an object. Using a lens this object is imaged onto a screen (Fig. 7.10). The measurement of the object and image distances (s_o and s_i) makes it possible to calculate the focal length using Eq. (7.18):

$$\frac{1}{f} = \frac{1}{s_i} + \frac{1}{s_o}.$$

This technique does not work when using objective lenses with focal lengths in the mm region because the distances get too small. In this case an illuminated object micrometer is used. This is projected onto the screen which is in a distance s_i of several 10 cm to the objective. By measuring the image size y_i the focal length can be calculated if the object size y_o is known. Since the object is practically in the focal plane of the lens ($s_0 \approx f$) Eq. (7.19) yields for the focal length:

$$f = s_i \frac{y_o}{y_i}. \tag{8.3}$$

The focal length of an objective can also be measured using a laser beam. To do so the laser beam is send through the objective and the expanded light cone is observed at a certain distance. In Eq. (8.3) y_o and y_i denote the diameter of the laser beam and the light cone at distance s_i, respectively. The result of this method is not very precise since the diameter of a laser beam can only be estimated without having performed a photometric measurement.

8.2.2
Experiment 6: Adjusting Lenses

When adjusting lenses these sometimes have to be rotated or set off parallel. To understand the occurring effects the following experiments are presented which are performed using a laser beam and an illuminated object. A rotation of the lens leaves the image position unchanged (Fig. 8.3); the lens aberrations are increased though. A parallel offset of the lens equally shifts the image. This is also true for the focus of a laser beam. Knowing these effects the adjustment of lenses becomes easier.

8.2.3
Experiment 7: Adjusting a Spatial Filter

A typical spatial filter works with an objective 40× with a focal length of $f = 4$ mm (= normed tube length 160 mm/magnification 40). Equation (7.20) yields for a He–Ne laser with a radius of $w = 0.4$ mm and a diameter in its focal point of

$$d = 2\frac{f\lambda}{\pi w} \approx 4 \text{ μm}.$$

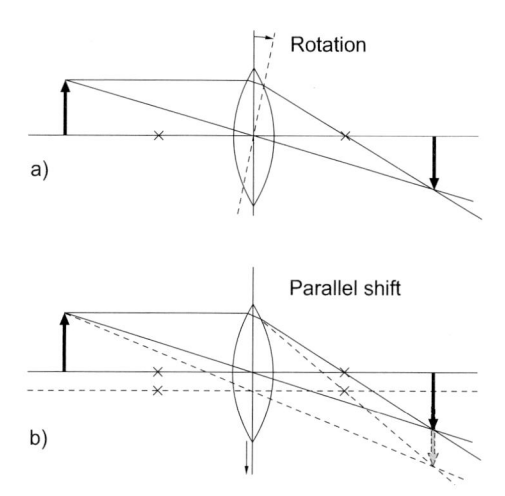

Fig. 8.3 Effect of lens movements on the image: (a) rotating the lens does not change the image position and (b) parallel shift moves the image in the same direction.

The diameter of the first Fresnel zone, d, is according to Eq. (7.21):

$$d = 2\sqrt{f\lambda} \approx 100 \text{ µm}.$$

Accordingly an aperture with a diameter of 40 µm is suitable for using it in a spatial filter.

The following procedure is favorable when adjusting a spatial filter: first thing to achieve is to transmit some light through the aperture. If this is done the aperture is adjusted perpendicular to the beam such that the brightness becomes maximal. Usually circular interferences appear which means that the beam diameter is too large. The aperture is now moved parallel to the beam in the direction where brightness increases while still adjusting the aperture perpendicular to the beam until the optimal position of the aperture is reached. If the expanded light cone is not symmetrical to the optical axis the spatial filter has to be set off parallel; a rotation yields no improvement.

If a divergent wave is suitable for the holographic purpose the beam emerging from the spatial filter can be used directly. If parallel light is needed a convex lens with large diameter according to Fig. 7.12 is used. Note that pulse lasers require special spatial filters. The focus diameter must not be too small to prevent electrical flash overs.

8.3
Experiments on Diffraction and Interference

8.3.1
Experiment 8: Diffraction at an Edge or Slit

Edge: If light falls onto the edge of a blade no sharp shadow appears. Interference fringes are formed at the shadows' edge due to diffraction. Figure 8.4 shows a setup for observation of this phenomenon where the diffraction pattern is magnified by a lens with $f = 20$ mm. A slightly expanded laser beam can be used as a light source.

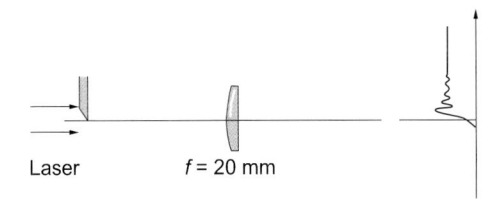

Fig. 8.4 Diffraction of light at an edge. The lens magnifies the diffraction pattern.

Slit: During illumination of a slit with coherent light diffraction occurs and several intensity maxima and minima are created (Fig. 8.5). For an introductory experiment a slit with a width of 0.2 to 0.4 mm is sufficient. To increase the diffraction angles a telescope made of two lenses is used. For example, using lenses with $f = 160$ and 20 mm, see Fig. 8.5, the resulting angle magnification is 8. If the slit is replaced by a wire of identical dimensions the resulting diffraction pattern is the same.

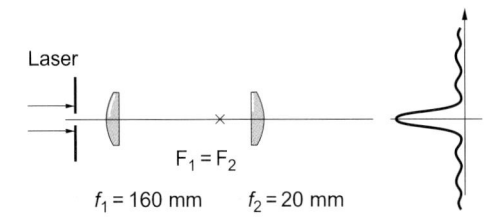

Fig. 8.5 Diffraction of light at a slit. Both lenses form a telescope with an angle magnification of $v = f_1/f_2$.

The experiment can also be done with circular apertures with a diameter of some 0.1 mm.

8.3.2
Experiment 9a: Interferences at a Glass Plate

A divergent beam is created using a lens; a part of the light is reflected by inserting a glass plate (Fig. 8.6). The light falling onto the screen shows noticeable interference fringes which can also be observed weaker in the transmitted light. The formation of the fringes is due to the light being reflected at the front and backside interfaces of the plate by about 4%. Both waves interfere and depending on the phase difference a maximum or minimum of the intensity occurs. The irregular structure of the fringes represents the variations in the thickness of the plate. At the Brewster angle the reflection and therefore the interferences disappear.

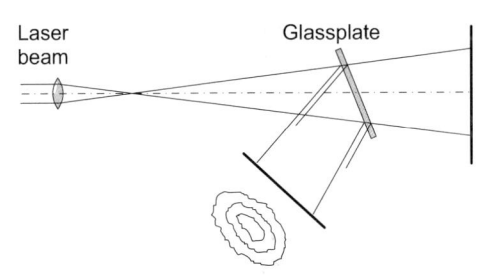

Fig. 8.6 Interferences at a glass plate. The contrast is larger in reflection than in transmission.

8.3.3
Experiment 9b: Newton Rings and "Index-Matching"

Similar interferences can be achieved with a lens with small curvature radius which is placed on a glass plate (Fig. 8.7). The superposition of the spherical surface of the lens and the planar surface of the plate reflected waves creates circular interference fringes which represent a height profile of the lens. These structures are called "Newton rings;" in the described experiment they are clearly visible.

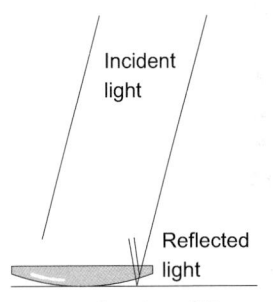

Fig. 8.7 Creation of Newton rings which represent the thickness of the air gap.

The same phenomenon is observable at the air gap between a holographic film and a glass plate. Newton rings can be prevented by filling the air gap with a liquid that has almost the same refraction index as the glass, e.g., turpentine. In holography this technique to prevent undesirable interferences is called "index-matching" (Section 14.2).

8.3.4
Experiment 10: Diffraction of a Grating

Several slits placed next to each other create a diffraction grating. If a grating is placed in a laser beam several sharp maxima appear to the left and the right of the beam. The deflection angle β of the diffraction orders is given by the grating constant d_g or the spatial frequency $\sigma = 1/d_g$ according to Eq. (2.35). For perpendicular incidence ($\sin \alpha = 0$) the following results:

$$\sin \beta = \frac{N \cdot \lambda}{d_g}.$$

For the beam of a He–Ne laser ($\lambda = 0.633$ µm) and a grating constant of $d_g = 2$ µm the angle for the first diffraction order ($N = \pm 1$) is $\beta = 18°$. For a sinusoidal grating no further diffraction orders exist; for nonsinusoidal gratings the appearance of orders with $|N| > 1$ can be explained by the fact that larger diffraction angles also mean higher spatial frequencies in the grating which have twice, triple, fourfold the main frequency. (In terms of mathematics this means that a nonsinusoidal grating is described by a Fourier series as a sum of sinusoidal gratings.) Using a grating the wavelength of a laser can be determined by measuring the diffraction angle. For precise measurements grating spectrometers can be used.

Interesting is the investigation of different light sources, e.g., gas discharge lamps and filament lamps. The observation of the spectra is done by illuminating a slit with a width of a few mm. The slit is observed from a distance of about 1 m or more by placing the grating right in front of the eye (Fig. 8.8). If the grating lines are oriented parallel to the slit spectral lines or a continuous spectrum appears left and right. The diffraction angles and wavelengths can then be determined by placing a scale in the plane of the lamp perpendicular to the observation axis.

If the discharge tube of a He–Ne laser is accessible it is worth observing it through a grating perpendicular to the discharge. Several spectral lines and bands become visible which are mainly caused by the neon. These are spontaneous emissions of light causing loss in the lasing process. In the axial direction the beam is monochromatic.

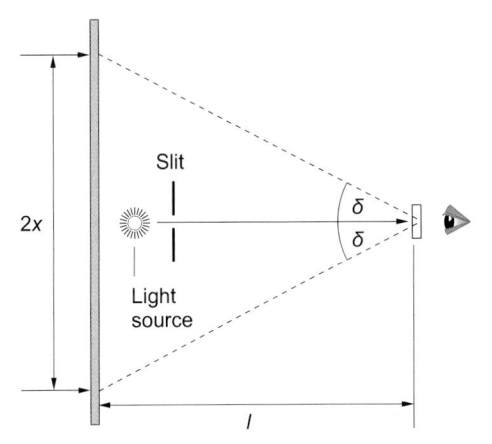

Fig. 8.8 Simple setup on grating diffraction. The diffraction spectrum coincides nearly with the plane of the scale.

8.3.5
Experiment 11: Divergence of the Laser Beam

Laser radiation propagates divergent according to Fig. 7.9. The divergence angle for common commercial lasers for holography is in the region of mrad ($= 10^{-3}$); a more precise value for a He–Ne laser with a radius of $w_0 = 0.4$ mm is

$$\Theta = \frac{\lambda}{\pi \cdot w_0} = 0.5 \times 10^{-3}.$$

Therefore, the beam expands by 1 mm after every meter. The expansion can be made visible by letting the beam propagate several 10 m. If the room is not large enough the beam can be reflected back and forth by using mirrors. The divergence is determined from the distance l to the laser and the beam radius at this point ($\Theta = w/l$). The result is not very exact though because the value for the radius is for the point at $1e^2$ (=13.5%), which is difficult to find visually.

The divergence affects the beam expansion by an objective or a lens. The opening of the expanded beam cone increases according to the laws of geometrical optics with the beam diameter at the lens. If the lens is placed directly in the exit of a He–Ne laser the beam diameter is about 0.8 mm and the result is a narrow expanded cone. If the lens is moved away 4 m the beam diameter is 4.8 mm. Accordingly the cone gets larger. Large distances between laser and lens are favorable for the beam expansion, also because the beam is "cleaned" by long paths, similar to a spatial filter.

8.3.6
Experiment 12: Optical Filtering

Figure 8.9 shows a simple experiment for demonstrating the optical filtering (or also for the imaging process in microscopes). A grating with a slit distance of 0.1 mm or less serves as an object which is illuminated with a parallel laser beam. Several diffraction orders appear at relatively small angle differences of 6×10^{-3} which pass a lens with $f = 10$ cm. The radiation of the diffraction orders is focused in the focal plane; for each order a focus appears about 0.6 mm beneath the zeroth order. This means, in terms of mathematics, that in the focal plane the "Fourier transform" of the object is created, i.e., the spatial frequency spectrum. The zeroth order describes the original laser beam, the first a pure sinusoidal grating with fixed spatial frequency. Higher orders characterize the deviation from the sinusoidal grating; they lie at twice, triple, or fourfold the spatial frequency.

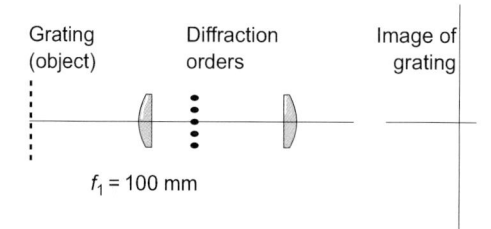

Fig. 8.9 Experiment on spatial filtering. Different diffraction orders appear in the focal plane representing the spatial frequency spectrum. By inserting apertures certain spatial frequencies (diffraction orders) can be suppressed. This is noticeable in the image created by the second lens.

By the use of a second lens the diffraction grating is imaged onto a screen. If several diffraction orders are able to pass the first lens a regular image can appear. In the experiment it shall be determined how the image behaves when diffraction orders are eliminated, e.g., by adding small apertures in the focal plane of the first lens. If only the zeroth order is permitted not an image of the grating but of the original laser beam itself is created. Adding the two first orders creates a sinusoidal grating. It gets interesting when only the two second orders are imaged. They represent the double spatial frequency and thus a grating with half the grating constant. The image therefore consists of a grating which is twice as tight as the original. In the described experiment particular frequencies have been filtered out after segmenting the object in different spatial frequencies.

8.3.7
Experiment 13: Granulation of Laser Radiation

Laser radiation differs from normal radiation by the fact that a granulation occurs during observation. This effect which has two causes results from the coherence of the radiation. One cause is the object from which the light is reflected. During reflection phase differences are created at the surface which cause granulation like interferences; these are called "speckles" (Section 5.4). Secondly a diffraction of the light at the pupil of the observer's eye occurs which creates a granulation on the retina. Both effects superpose. In the following some simple experiments for understanding the granulation will be described.

To investigate the granulation the beam of a He–Ne laser is expanded to about 20 cm and aimed onto a screen. A strong granulation is visible in reflection as well as in transmission. It remains in focus independent of the observer wearing glasses or not – even with the eye very close to the screen. The granulation only changes if the surface structure is changed.

The granulation is averaged out when the reflecting body is moved fast since the eye is too slow to follow the movement of the pattern. This can be observed, e.g., in light scattered from milk where no granulation is visible because the thermal movement of the scattering particles is too fast. However, by adding some acid the milk coagulates and scattering is observed. Therefore their velocity is decreased and the granulation becomes visible. The same effect occurs if a screen is alternately moved fast and slow.

For a long-sighted eye the focus lies behind the retina. For this reason the granulation moves in the same direction when moving the head. For a short-sighted eye the direction is inverted since the focus is in front of the retina. For a normal eye an irregular movement with no preferred direction is visible. Thus the observation of granulation is a simple possibility of checking for eye defects and testing the correct choice of glasses.

The influence of the eye pupil on the granulation can be analyzed by placing a small pinhole in front of the eye. A decrease of its diameter causes an increase of the size of granulation because the diffraction angle increases. The influence of the diffraction at the pupil is also visible for normal light sources at a large distance. In this case the light propagates almost parallel and is partially coherent so that in the light of far away lamps sometimes granulation becomes visible.

8.4
Measurements with Interferometers

8.4.1
Experiment 14: Setup of a Michelson Interferometer

In interferometers the light wave that is to be analyzed is divided into two separate waves which are changed geometrically and are superposed again afterward. Laser interferometers after Michelson are used in holography to measure the coherence length and the stability of the optical setup. The laser beam hits a beam splitter (Fig. 8.10) and is divided into two waves of the same intensity: the passing wave 1 and a perpendicularly traveling wave 2. The latter is reflected by a plane reference mirror and wave 1 is reflected by a movable mirror into itself. Part of the reflected wave is guided onto a screen by beam splitter T. Since the two waves are always tilted a little bit in practice a system of fringes appears on the screen due to interference. If the movable mirror is shifted by a distance Δl the interference pattern moves by m fringes according to the following equation:

$$\Delta l = m\frac{\lambda}{2}. \tag{8.4}$$

By counting the shifted fringes the distance Δl can be determined with high precision.

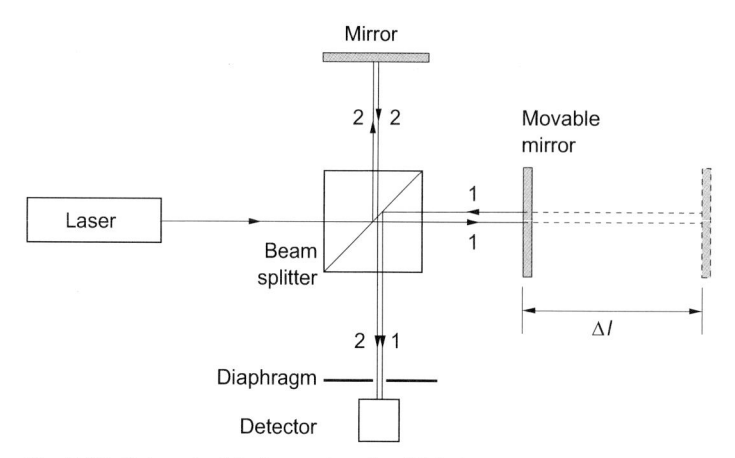

Fig. 8.10 Setup of a interferometer after Michelson.

The interferometer can be setup using optical components, but it is advised to place the movable mirror onto an optical bench. The laser is expanded in front of the interferometer by a lens or a telescope setup. Using a lens results in spherical waves such that a circular interference pattern is created. The mirrors should be equipped with fine adjustment. The adjustment of the setup is done by superposing the waves on the screen. Here little defects or dust particles in the laser beam can be very helpful. The sensitivity of the interferometer can be demonstrated by heating the air under one of the beams with the hand or a lighter which causes a shift of the fringes.

8.4.2
Experiment 15: Measuring Coherence Length

In Sections 7.1 and 7.2 it was noticed that laser radiation has a limited coherence length due to the mode structure. The coherence length is several meters if only one longitudinal mode exists which is achieved by using an etalon in the resonator. The measurement of the coherence length in this case is difficult due to practical reasons. Instead a spectrum analyzer is used which represents an automatic Fabry–Perot etalon. The coherence length is determined by measuring the line width. In holography this method is used to adjust ruby or argon lasers in longitudinal single-mode operation.

Holographic lasers which emit several longitudinal modes have essentially shorter coherence lengths in the 10 cm region, which can be determined with a Michelson interferometer. For the measurement, e.g., of a He–Ne laser, a Michelson interferometer according to Fig. 8.10 is setup. When shifting the movable mirror the visibility of the interference gradually decreases. This can be observed with the naked eye or measured with a line camera. The visibility is defined according to Eq. (2.12a). During the measurement the movable mirror has to be readjusted after each shift. For a He–Ne laser of about 1 m length the resulting curve is shown in Fig. 8.11. The coherence length is given by the shift for which the visibility drops to $1/\sqrt{2} = 0.707$. The periodicity in Fig. 8.11 is because the individual longitudinal modes lead to periodical beats.

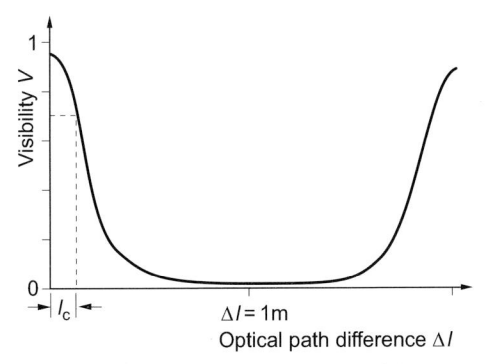

Fig. 8.11 Measurement of the coherence length (He–Ne laser) by moving a mirror of a Michelson interferometer by Δl. The coherence length is given by the distance $\Delta l = l_c$ for which the visibility drops to 0.7.

8.4.3
Experiment 16: Investigation of Stability

During the holographic recording it has to be ensured that the interference fringes move a few 10 nm at the most so that no decrease in the diffraction efficiency occurs. The required stability of the holographic table can be checked with a Michelson interferometer. To enhance the sensitivity the distance of the mirrors from the beam splitter should be large. The interference fringes are projected onto a wall and the stability of the fringes can be observed under different conditions, e.g., with people walking by, when switching on the air conditioner, during air movements in the room and the like. An interesting experiment is placing the laser beneath the optical table. It then can be observed that vibrations at the laser have no effect on the fringes since they occur in both partial waves and compensate each other.

8.5
Production of Gratings and Simple Holograms

This section contains suggestions for experiments with a simple setup on single-beam holography which can be used as introductory laboratory experiments. No special demands are put on the stability of the table since components for object and reference waves are mechanically connected. In the first experiment a holographic grating is produced, in the second a simple hologram.

8.5.1
Experiment 17: Production of Diffraction Gratings

Setup: Transmission

A grating can be created by the superposition of two waves. At a simple setup both waves are extracted from an expanded laser beam by two mirrors, see Fig. 8.12. The beam of a He–Ne laser is expanded by a microscope objective $40\times$ with a focal length of $f = 4$ mm. The objective is part of a spatial filter. In a distance of about 40 cm an expansion of the beam diameter by a factor 100 occurs and up to a diameter of 10 cm. Using a convex lens the beam can be parallelized again. The distance from the objective (more precisely from the focal point) is chosen so that it is the same as the focal length of the lens. The diameter of the lens needs to be quite large for the expanded wave to pass.

In the path of the beam a mirror is placed at $45°$ which deflects the beam in the perpendicular direction (Fig. 8.12). A second mirror is leaned against the first one so that it reflects half of the beam at a slightly smaller angle.

8.5 Production of Gratings and Simple Holograms

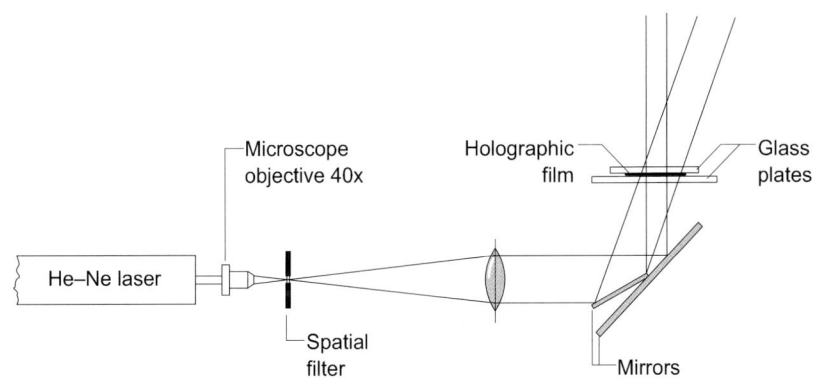

Fig. 8.12 Experimental setup for the production of diffraction gratings.

The beams superpose and create a plane interference system in the form of a spatial light grating.

A glass plate is positioned horizontally using a table-like setup. It serves as a carrier for the holographic film which is fastened into a flat position by a second plate.

The setup is placed onto a vibration isolated table which in the simplest case consists of a 50 to 100 kg concrete plate which is placed onto some bicycle or motorcycle tubes. The air pressure is adjusted so that the system can oscillate with 1 to 2 Hz. This can be easily verified by giving the table a push. A simple optical shutter consists of a cardboard hanging from the ceiling by two strings.

Setup: Reflection Grating

The interference fringes lie on the bisector of the angle between the two beams. In the case described above the fringes are oriented almost perpendicular to the film plane and a transmission grating is created (Chapter 2). The experiment can be modified to create a reflection grating. The second mirror is removed and placed onto the holographic film instead, possibly slightly tilted, which deflects a second beam onto the film from the top. The beams travel in opposite directions; the bisector and therefore the grating planes are now almost parallel to the film plane. The working principle of a reflection grating is similar to a selective mirror which only reflects a specific wavelength at the Bragg angle. The experiment can also be used to produce a holographic concave mirror by replacing the second flat mirror on the photographic layer by a concave one (Chapter 16).

8.5.2
Exposure and Developing

Suitable are holographic films or plates PFG 01 by Slavich or Fuji F-HL with a sensitivity of 75 µJ/cm² at an optical density of 2.0. In practice good results are achieved using a 5 mW laser with an exposure time of a few seconds. After exposure the film is developed, bleached, and dried. All the steps are presented in the experiments description in Section 8.6.

8.5.3
Experiment 18: White Light Hologram

The above-described experiment can be easily modified to produce holograms. The new setup is shown in Fig. 8.13. Since parallel light is not required the large lens from Fig. 8.12 can be omitted. Thus the beam diameter on the holographic film is slightly larger. The object is placed onto the glass plate above the holographic film. The back-scattered light from the object is the object wave; the light coming from beneath is the reference wave. Since both waves come from different sides a white light hologram is created in reflection. The deflecting mirror in Fig. 8.13 enhances the stability since the object sits stationary on top of the holographic layer.

To observe the holographic image it can be illuminated with a spot-like light source, e.g., a 12-V tungsten lamp. It is beneficial to blacken the backside of the hologram with spray color to absorb the nondiffracted light. Since the direction of the illumination should coincide with the reference wave it is favorable to place the mirror in Fig. 8.13 at an angle of 30° to 45°. With a 5 mW laser an exposure time of about 10 s is needed. To create an intense object wave objects with a high backscattering should be chosen. It is also advised to put the emulsion side of the film toward the object.

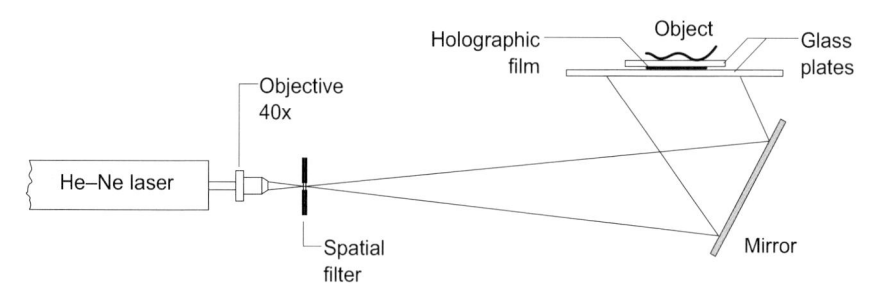

Fig. 8.13 Setup for single-beam holography for the creation of white light holograms.

8.5.4
Experiment 19: Transmission Hologram

Experiment 18 describes the production of white light holograms; reference and object waves fall onto the film from opposite sides. If both waves fall from the same side onto the layer a transmission hologram is created. In a simple setup the object and the holographic film are placed in the expanded beam such that the film is hit by the object and reference waves from the same side. The film is put into a holder between two glass plates.

After inserting the film one should wait at least 10 min since the film still moves a little bit. Although the stability of the setup in Fig. 8.13 is higher a waiting period is required, too. An undesired movement of the film can be noticed later as regions with no information which appear dark in the reconstruction. The produced transmission holograms only deliver sharp images when illuminated with a laser.

8.6
Experiments in the Darkroom

The following section illustrates three simple experiments on the developing and bleaching of holographic films. The exposure creates a so-called latent image on the holographic film which gets visible during the developing; the result is an amplitude hologram. To enhance the diffraction efficiency it is transformed into a phase hologram by "bleaching." This procedure can cause a change in the thickness of the holographic layer which again can cause a color shift in the reconstructed images. A more extensive description of holographic layers and their treatment is given in Chapter 14.

8.6.1
Experiment 20: Developing

During the developing AgBr crystals that were hit by light quantum are reduced to Ag. For laboratory training on holography standard commercial developers are suitable, e.g., Kodak D 19 or Dokumol for document films, but also CWC2 and other special holography developers may be used (Section 14.2).

Using a simple setup as shown in Fig. 8.14 the darkening of a film after developing can be measured. Reflection holograms are developed until an optical density of $D \approx 2$ is reached, i.e., the developed film has a transparency of $10^{-D} = 10^{-2} = 1\%$. For transmission holograms the darkening should be a little bit less ($D \approx 1$). After the developing the film is rinsed under tap water for 5 min, and afterward with deionized water for 5 min. The result is an amplitude hologram. The diffraction efficiency of this hologram type is so low

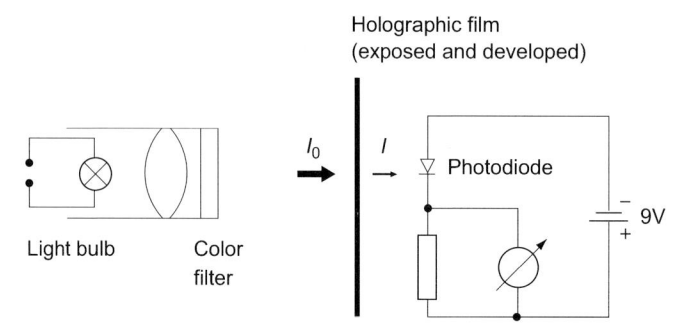

Fig. 8.14 Simple setup to measure optical density $D = \log I_0/I$ when developing holograms.

that it is usually transformed into a phase hologram by bleaching. However for demonstration purposes the amplitude hologram may be desensitized by fixative or longer rinsing.

8.6.2
Experiment 21: Solving Bleach Bath

To transform an amplitude hologram into a phase hologram the exposed Ag can be removed from the gelatin layer by a so-called solving bleach; the fixative can be omitted then. The following recipe for a dichromate bath can be used: 1 liter of deionized water, 1 to 7 g of potassium dichromate, and 1 to 7 ml of concentrated sulfuric acid. The hologram is left in this solution until all black areas are clear. After bleaching the hologram is desensitized in water for another 15 min and after that it is washed in relaxed water or alcohol. The hologram is not visible until after drying (e.g., with a blow dryer). The yellowish coloring by the dichromate may be removed by a cleaning bath after bleaching (recipe for the concentrated solution: 50 g sodium sulfite, 1 g sodium hydroxide, 1 liter of deionized water. For later use: about 20 ml on 0.5 l water). The thickness of the holographic layer is reduced due to the bleaching process. This causes the Bragg reflection to shift to shorter wavelengths since in white light holograms the interference planes are almost parallel to the layer. When using a He–Ne laser the images appear green. In transmission holograms the interference fringes are crosswise to the film plane so that the fringe distance is not changed by a shrinkage of the layer. In this case no color shifts are observable.

8.6.3
Experiment 22: Rehalogenizing Bleach Bath

A shrinkage of the layer thickness can be prevented by using a so-called rehalogenizing bleach. In this process the Ag is not removed but converted into a different chemical compound. Thus for white light holograms no or only minor changes in color appear. The solution prepared according to the recipe described above can be transformed into a rehalogenizing bleach bath by adding 4 to 16 g of KBr. As a demonstration two similar holograms may be bleached different so that a green and a red image is created. Both holograms can be created simultaneously by placing two films on top of each other.

Tab. 8.1 Developing and bleaching of holograms. (The cleaning bath is not necessary for all bleaching baths)

	Baths	Process	Minutes
1.	Developer	Creation of an amplitude hologram	0.5–2
2.	Tap water	Removal of developer, Desens.	5
3.	Dist. water	Removal of tap water	1
4.	Bleaching bath	Transformation into phase hologram	ca. 1*
4a.	Dist. water	Removal of bleaching bath	0.1
4b.	Clearing bath	Decoloring (if needed)	2
5.	Dist. water	Desensitizing	15
6.	Rel. water	Washing	1
7.	Drying	Hologram becomes visible	10

*Bleaching until completely transparent

Problems

Problem 8.1 A laser wave of a He–Ne laser is oscillating in the y-direction. (Linear polarized, see Fig. 8.1). The plane of polarization shall be rotated about $45°$ with respect to the original direction. What action has to be taken and how could one analyze that the problem is solved in the right way?

Problem 8.2 A holographic diffraction grating has a grating constant corresponding to 800 lines/ mm. Inspecting the spectrum of a white light point source, at some angle you see a purple color, red (600 nm), and blue (400 nm) overlap, obviously. Of what diffraction order is the red and the blue light. (Experiment 10)

Problem 8.3 A Michelson setup (Fig. 8.10) is build. The visibility of a laser beam decreased to 70%, if the signal mirror was moved about $\Delta l = 15$ cm. Calculate the bandwidth Δf and $\Delta \lambda$ of the He–Ne laser used and the number of longitudinal modes ($\lambda = 632.8$ nm, length of laser $L = 75$ cm) (Experiment 14).

Problem 8.4 Observing a point-like white light source perpendicular to a holographic grating, the virtual image of the red light (600 nm) is seen in the direction of $45°$ with respect to the point source. What angle should be between the two mirrors shown in Fig. 8.12 to make such a grating, using a He–Ne- laser ($l = 632.8$ nm) or an Ar laser ($l = 514$ nm)?.

Problem 8.5 To make a single beam transmission hologram (Experiment 19) using a setup given in Fig. 8.13, where should the object be placed?

9
Experimental Setups for Single-Beam Holography

9.1
Setups for Reflection Holograms

Reflection holograms that can be reconstructed with white light require a deeper theoretical understanding of holography. On the other hand, their production is very easy especially with the single-beam process which is done without beam splitting.

9.1.1
Experimental Setups

Figure 9.1 illustrates a simple experimental setup. The photographic plate is placed into the expanded beam at an angle of about 30°. The object is placed behind the plate seen from the laser. The reference wave serves also as the illumination wave after passing the photo plate. The path difference between the reference and object waves is therefore twice the distance between the object and photo plate. To stay inside the coherence length of the laser (about 20 cm to 25 cm for lasers around 10 mW) the object should not be placed more than 10 cm away from the plate.

This recording technique with the photo plate placed at an angle in the

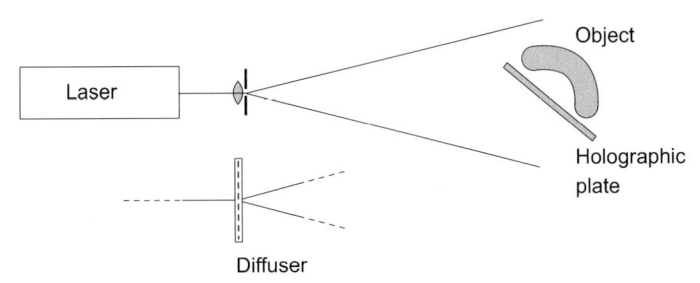

Fig. 9.1 Simple single-beam setup for white light reflection holograms with the expanding system or diffusing screen.

Holography: A Practical Approach. Gerhard K. Ackermann and Jürgen Eichler
Copyright © 2007 WILEY-VCH Verlag GmbH & Co. KGaA, Weinheim
ISBN: 978-3-527-40663-0

beam is called "off-axis holography." This method dates back to Leith and Upatnieks (Section 3.2). Contrary to the "inline-holography" developed by D. Gabor (Section 3.1) many advantages result for the reconstruction of the image: after reflection the required wave (identical to the reference wave needed for recording) propagates in a different direction than the reconstructed object wave.

An observer sees the reconstructed virtual image and is not dazzled by the reconstruction wave itself.

9.1.2
Index Matching of Holographic Films

Instead of holographic plates for many applications films are used. Holographic films are less expensive and easy to handle. Slavich produces sheets of 20 cm × 30 cm size and different rolls between 10 m and 20 m total length. Details on how to process the Slavich holographic material is discussed in Chapter 13.

Films are fixed between two flat glass plates. Because of the many interfaces and nonperfect glass surface unwanted interference patterns are created by multireflection of the laser light during exposure. Using antireflection glasses the problem is reduced. But the intensity of the polarized radiation is reduced, too [27]. There are two procedures to overcome such interference patterns.

To suppress all unwanted reflections an "index-matching" can be applied. Within this technique the differences of the refractive indexes of glasses, film and air are bridged by a fluid of almost the same refractive index as glass and film ($n = 1.5$). The applied fluid between the glass plates should not be volatile (no alcohol) in order to prevent small amounts of movement due to evaporation of fluid during exposure. Of course, the refractive index should almost match that of the glass and film. Glycerol, liquid paraffin, carbon tetrachloride, xylene, or paint thinner (white spirit) are used [27]. Most of them do not smell very good and some are harmful to health.

Saxby [27] recommends white spirit because of its high surface tension. Therefore, one glass is necessary, only. In a first step the glass plate is cleaned thoroughly; then some drops of white spirit are deposited in the center of the glass plate and bowing the film a little bit (emulsion side up) it touches the fluid in the center first. The surplus liquid is dabbed of with absorbent paper and the setup rolled over once with a roller; the so prepared film can be placed straightaway in the plate holder. Due to the frequent changes of the emulsions and especially the carrier material pilot tests are indispensable to find the right liquid to optimize the procedure.

9.1.3
Setups without Index Matching

Simple setups exist for which it is almost unnecessary to apply the "index-matching." Two setups will be shown exemplary which are easy to realize and are well tried (Figs. 9.1 and 9.2). In both cases the laser beam impinges on the plate at the Brewster angle. Light with a specific linear polarization that impinges at the Brewster angle (for glass ≈ 56°) on a glass plate is transmitted completely and not reflected. The polarization has to be aligned as shown in Fig. 8.2. If the electrical field strength lies in the plane of the glass the light is partially reflected and undesired reflections and interferences occur at the interfaces.

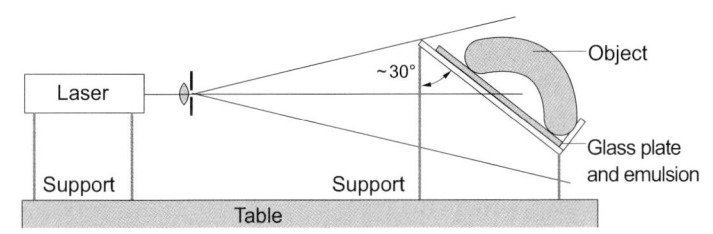

Fig. 9.2 Single Beam setup for white light reflection holograms using Brewster's angle.

The direction of the vector of the electrical field strength and therefore the polarization direction can be checked by using a glass plate (Fig. 8.2). The glass plate is slowly rotated inside an expanded laser beam around a perpendicular axis into the Brewster angle. If the polarization direction of the laser is adjusted "right" then a noticeable minimum in the reflected light can be observed in a room with dimmed light. If the decrease of the reflected light does not appear then the polarization direction of the laser is rotated by 90°. This can be checked by turning the glass plate around a horizontal axis. In order to adjust the setup either the laser needs to be rotated or the polarization direction by 90° using a $\lambda/2$-plate. The latter though causes a small loss in intensity.

9.1.4
Intensity Loss at Brewster-Angle Setting

The method of using the Brewster angle to eliminate unwanted multireflections can lead to a loss in visibility of the hologram. The laser light that is reflected by the object during recording is mostly completely depolarized which can be checked with a polarizing filter. However if the surface of the object is metallic reflecting the polarization direction is preserved. The setup using the Brewster angle causes that the polarization directions of object and reference waves are not parallel but enclose the Brewster angle. Now only the partial amplitude $A = R \cos \varphi$ of the reference wave interferes with the object wave. In the equation R denotes the amplitude of the reference wave and φ the Brewster angle. At an angle of nearly $60°$ A is only half as large as R.

9.1.5
Vacuum Film Support

The most elegant solution for film support is a vacuum film frame. Such a support is shown in Fig. 9.3. A glass plate is placed in an aluminum frame with a cut channel of a few tenth of a millimeter in depth. The rectangular area is exactly of the film size. The film is inserted in this frame and the vacuum pump is started. This setup may also be used with the index matching technique. The exposure should be started after some minutes of settlement.

If the frame is used for transmission holograms only, the backside can be full aluminum and the vacuum channel can be redesigned to cover the whole backside of the frame.

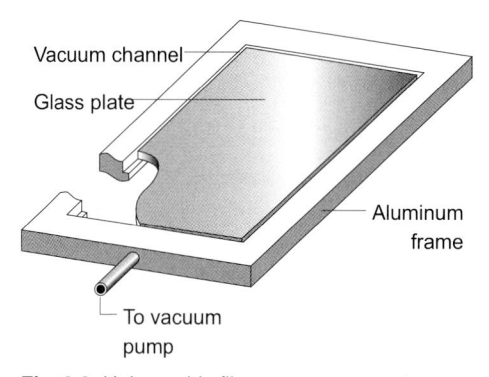

Fig. 9.3 Holographic film vacuum support.

9.1.6
Simple Single-Beam Setups

The most simple arrangement of a single-beam setup only needs a frosted glass which is placed at the position of the expanding system in Fig. 9.1 and expands the beam sufficiently. By doing so a hologram can be produced very fast but with the drawback that it has a grainy structure caused by the granulation of the laser radiation (speckles).

Figure 8.13 sketches a setup in which the photo plate or the film lies flat on a glass plate. The mirror is placed in a way that near-axial beams are incident on the glass plate and the film at the Brewster angle. For this setup also an index matching could be performed since the film lies horizontally on a glass plate.

A very simple setup which uses the Brewster angle is shown in Fig. 9.2. To support the objects a slim strip of glass or wood is glued to the glass plate (e.g., using easy to handle hot glue). This Brewster setup is quite rigid and capable of imaging larger objects.

9.1.7
Holographic Table

In single-beam setups for producing reflection holograms low demands are put on the stability of the table. The overall arrangement only contains a few optical components and is usually so compact that vibrations affect it only as a whole. These do not have a negative effect on the diffraction efficiency. More problematic are vibrations of individual optical components with respect to each other. They cause temporal varying phase shifts of the interfering waves and therefore a deletion of the already recorded information. Thus the diffraction efficiency is decreased. Consequently the holograms remain weak even with long exposure times or are not makeable at all.

Vibrations can be avoided by vibration isolated tables. Different solutions to this problem are described to extend in Section 7.7.

9.1.8
Visibility

Single-beam reflection setups have one serious disadvantage. The intensity ratio of object wave to reference wave cannot be changed by easy means. The visibility can decrease quite fast for bad reflecting objects. The limit for interference fringes being visible is around $V = 0.2$ (for calculation of V see Section 2.1.3). This limit by experience should be exceeded significantly. To work with a visibility below $K = 0.5$ is not advisable. This value corresponds to a ratio of object wave to reference wave of about 1:15.

9.2
Setups for Transmission Holograms

Figure 9.4 shows two setups for the recording of transmission holograms. From the figure it is apparent that object and reference waves hit the photographic plate from the same side. Within certain limits it is possible to adjust the intensity ratio during recording by reducing that part representing the reference wave without effecting the illumination of the object.

The single-beam transmission setups most of the time do not allow a positioning using the Brewster angle because object and reference waves hit the photo plate from very different directions.

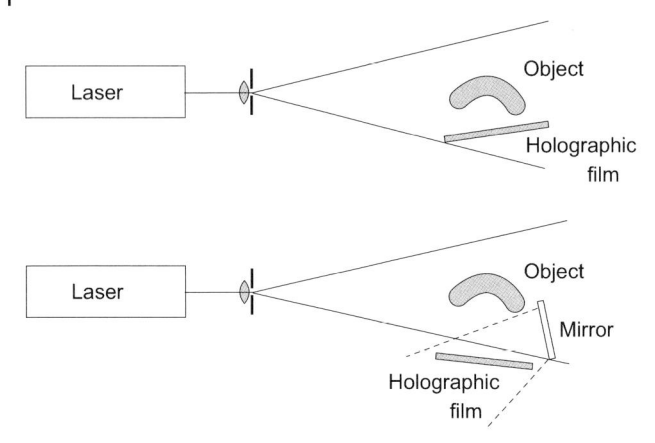

Fig. 9.4 Single-beam setup for transmission holograms.

These setups are good for first experiments on producing transmission holograms because few optical components are required. For master holograms which are to be used to make holograms of holographic images these single-beam setups are hardly suitable, at best for the production of simple "holographic optical elements" (HOE) or for first experiments on double exposure interferometry (Section 15.2).

A single-beam setup has the advantage of putting low demands on the stability of the table. The expansion system in transmission holograms can be replaced by a frosted glass, too. However, this simplification in setup leads to increased noise in the hologram due to the granulation (speckles) of the laser light.

Transmission holograms have a larger depth than white light holograms. Due to the way the arrangement is set up – with object and reference waves coming from the same side – the path difference between the waves can be kept small. Therefore the coherence length is not reached as quick. The images of the holograms are reconstructed using the monochromatic radiation of the recording laser.

When positioning the object one has to make sure that no shadow falls onto the photo plate. To check this a white cardboard is placed into the hologram holder which also checks the illumination by the reference wave. Before recording it should be checked that no reflections from the different components of the setup hit the photo plate. By skillful positioning of screens (e.g., black cardboard) this scattered light can be blocked.

To prepare for the recording the laser beam is blocked by a piece of cardboard or if available by a photographic shutter. The photo plate is placed into the holder in not too bright green laboratory light. Due to cost reasons most often films are used that are placed between two thin glass plates (thickness about 1 mm). The emulsion side of the plate or film can be determined by an aspirating test. The free glass or film side will mist if aspirated. The arrangement is then rolled under high pressure with a rubber roll so that all the air escaped from between the different layers. After that the glass plates are clamped on three sides. Subsequently this arrangement is placed into the holder with the emulsion side facing the object.

The exposure time depends on the laser used, the geometry, and the developer so that advices given here are general in nature. The developing should be carried on until an optical density of about 1. Optical density 1 means that the transmission of the photo foil is 10% of the starting value after processing. If this parameter is not exactly measurable during the process, the developing is done until exposed and unexposed areas (e.g., at edges) are darkened noticeably different. However the exposed areas are still transmittend.

To find out the correct exposure time it is a good idea to make test recordings on strips of the photographic material. A test series with 2 s, 4 s, 8 s (16 s) is recommended. The doubling of the exposure time is necessary to get distinct visible effects in processing the film. The given test series is suitable for a 10 mW laser, 40× expansion system, 4" × 5" photo foils, and about 1 m overall length of the setup. Decisive for the optimum exposure time is the brightness of the reconstructed image, its diffraction efficiency.

9.3
Experiments for Image Reconstruction

9.3.1
Reconstruction Angle

The reconstruction of the image is done by placing the hologram at the same position with the same orientation as during the recording. Figure 9.5 shows the image reconstruction for a transmission hologram. Light that is coming from the same direction as the reference wave must be used, with the same curvature (plane wave, spherical wave) and with the same wavelength (Chapter 6). It is not advisable to change the direction of the reconstruction wave since the diffraction efficiency quickly drops with varying angle. In this point there are not many possible variations.

9.3.2
Reconstruction Light Source for Transmission Holograms

When using white light for reconstruction a continuous spectrum instead of a sharp image is reconstructed in which only the contours of the object are indistinctly visible. The transmission hologram then acts like a diffraction grating

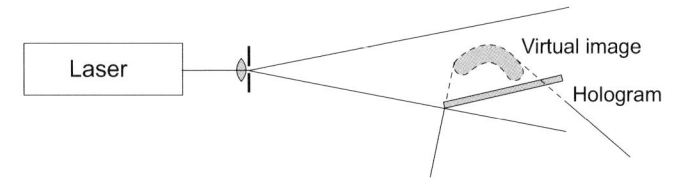

Fig. 9.5 Reconstruction of the image for a transmission hologram.

and reconstructs the object according to diffraction theory for different colors with different angles which lie close together for a continuous spectrum.

Another smearing effect can be caused by the light source not being point like (e.g., a frosted light bulb). The reconstructed image remains blurred even when a narrow band filter is used and the light almost resembles the one of the laser that was used during recording. With the laser that was used for the recording it is possible to reconstruct a pin-sharp image. If the hologram is cut up, each individual piece shows the complete object but only in that perspective that could be seen through this part of the original hologram. From that point of view each hologram piece contains the information of the complete object but not for all perspectives.

It is also possible to use other monochromatic light sources for reconstruction. With an argon laser instead of a He–Ne laser the object image can also be sharply reconstructed but at a different angle and geometrically smaller or larger depending on the wavelength being smaller or larger as during the recording (Eq. (5.5b)).

9.3.3
Reconstruction of the Real Image

According to the remarks given in Chapter 6 the reconstruction with the conjugated beam displays the real image. To do so the hologram is rotated by 180° and is illuminated from the opposite side with the same reference wave. If the reference wave is a plane wave the real pseudoscopic image of the object is reconstructed true-to-scale.

In the single-beam setup of Fig. 9.4 the reference beam is a spherical wave. According to Chapter 2 the conjugated wave not only has to come from the opposite direction but it has to have the curvature of a converging wave. But turning the hologram only changes the direction of the incoming light; the wave itself remains divergent. Therefore the image is reconstructed at a larger distance from the plate and it is magnified. Sometimes the magnified reconstruction is required when small objects are to be displayed (e.g., when creating master holograms, see Chapter 11 and Section 5.1.

9.3.4
Reconstruction of Reflection Holograms

The above-discussed experiments cannot always be successfully performed for so-called white light holograms, in this case white light reflection holograms. The reconstruction of the images in white light holograms can be performed using a common light bulb with clear glass body (point light source). The hologram reconstructs the image in that color which is given by the plane distances of the developed silver in the photo film (Bragg condition). Depending on the bleaching bath that was used and the caused shrinkage of the emulsion the wavelength of the reconstructed light for a He–Ne laser lies in the green to red spectral region. If the hologram is illuminated with monochromatic light with a wavelength not matching the distance of the just mentioned planes, the diffraction efficiency decreases.

If a reflection hologram is cut up, often not the whole object can be reconstructed by the individual parts. The reason for this lies in the setup itself. Since the object lies flat on the film during recording not all of the light scattered by the object can reach every part of the hologram. Therefore not all the parts can reconstruct the complete image.

9.3.5
Wavelength Shift

If a solving bleaching bath is used by which the developed silver is removed from the emulsion and a phase hologram is created, the emulsion layer shrinks a bit. This way the wavelength for reconstruction shifts in the case of reflection holograms since the grating constant has decreased. The amount of shift depends on the developer used.

There is a simple possibility of changing the thickness of the photosensitive layer by soaking it. Holding the hologram in water steam causes the photosensitive layer to become thicker by soaking with water and the color is shifted to longer wavelengths. Drying the films, e.g., in a heating oven, the reconstruction wavelength becomes shorter due to the shrinkage of the emulsion layer and the color changes from e.g., yellow–green to blue–green. However a permanent change of color cannot be achieved this way. Permanent color changes by soaking the photosensitive layer can be achieved by using bathes of triethanolamin of different concentrations. This aspect is discussed in more detail in Chapter 12.

9.3.6
Reconstruction of the Real Image

The reconstruction with the conjugated reference beam (by turning the hologram by 180°) shows the real pseudoscopic image which is now in front of the emulsion layer. This effect can be used to produce a white light reflection hologram with the image in front of the plate in one step. First, a pseudoscopic cast of the object is produced, e.g., a good reflecting plaster cast. From this a white light reflection hologram is produced as described above. The reconstructed real image of the conjugated beam is then again orthoscopic and lies in front of the photo plate.

9.4
Trouble Shooting

In this chapter many different setups for holographic exposures are presented. Here some problems are discussed which may arise during or after exposure. Unlike photography it may happen that after exposure and chemical development the plate turns black but after bleaching there is no image at all. In order to be sure that no image is stored one should let the plate dry before bleaching without using a blow dryer. The holographic plate is inspected in transmission with a spot light (even if it is a reflection hologram). The plate should be tilted in different directions to meet the direction of the reference beam. After a while some colors should be seen. This is an indication that a hologram can be expected. Even if no colors are seen the development process (bleaching) should be continued and the result inspected. There may be an image or not.

There are many reasons for not getting any image:

- The reason for "no hologram" may be that one of the components in the setup is loose. Some parts moved. Half a wavelength is enough to destroy the image stored. The setup has to be checked first.

- There may be no image because during completing the setup and doing necessary measurements one of the beams was blocked off and the light stop is still there.

- There may be no image because the intensity ratio is not correct. One should check the intensity of both beams again and see whether the object beam is too faint or even stronger than the reference beam. In both cases there will be almost no image.

- If no image can be found it could be that the developing created the complex form of silver. After developing and washing the plate for some minutes under full light in the darkroom the silver looks light brown-red in transmission instead of black. Then with solving bleach all information is erased. Rehalogenating bleach should only be used (e.g., PBU).

- There is no visible image in a white light hologram. If triethanolamin (TEA) is used to improve the sensitivity, it could be that the reconstruction wavelength moved far into the blue region and the image is not detectable. Just aspirating the emulsion side of the plate the emulsion will swell to some extent and the reconstruction wavelength will move into the green area or even to the red, depending on aspiration. The image can be seen then for the time the emulsion is wet. Less concentrated TEA should be used for the next holographic plate.

- There is an image but there are some black lines or areas where no picture is to be seen. In this case the object or the film support moved. Maybe the index matching fluid moved or evaporated and maybe the waiting time was too short. Also if hot glue was used to fix some parts it takes at least 10 min until the glue is hard enough. Hot glue fixing should be done well before exposure.

- After developing, bleaching, and drying there are many colored lines covering the whole plate. This is not a severe problem. The colored lines are due to the light entering the holographic plate from a side creating wavelength dependent interference lines because of multi reflection between the glass–air interfaces. If the sides are blackened or covered with opaque adhesive tape the lines will vanish. This can be done before or after exposure and developing.

- After developing and bleaching the image is very faint and may be covered with some deposit. In this case one should refresh the applied solutions.

Problems

Problem 9.1 To make a white light reflection hologram a single beam setup is used, similar to Fig. 9.1. Calculate the visibility for an object of 70% reflectivity for the wavelength of the laser. The transmittance of the holographic plate is 90%.

Problem 9.2 The transmission of a developed photo plate is 1/30 of the original light intensity I_o. Calculate the optical density.

Problem 9.3 For a single beam transmission hologram setup (Fig. 9.3, lower one) the intensity of the object wave (mirror removed) at the position of the holographic plate was $I_o = 4.5$ a.u. Both waves together have an intensity of $I_{r+o} = 20$ a.u. Calculate the visibility. (a.u. = arbitrary units)

Problem 9.4 To reconstruct the virtual and the real image of an object (Fig. 9.4, exposure: divergent reference wave, He–Ne laser, $\lambda = 632.8$ nm) an Ar laser is used ($\lambda = 514$ nm, same position and divergence). Calculate the lateral magnification assuming $z_c = z_r = 100$ cm. The distance of the object from the plate was 15 cm. Make a plausible assumption about the shrinkage factor m (transmission hologram!)

Part 3   Advanced Experiments and Materials

10
Experimental Setups for Split-Beam Holography

For the production of holograms two waves are used, an object wave and a reference wave. One speaks of "single-beam holography" if the expanded laser beam serves as the illumination and the reference wave at the same time. In the setups described in this chapter reference and illumination waves are used that were separated by a beam splitter. This so-called split-beam-holography will be described in the following.

10.1
Setups for Transmission Holograms

10.1.1
Experimental Setup

Figure 10.1 shows a typical setup for a transmission hologram [20]. The first beam splitter divides the beam into reference and illumination wave. These are then guided by different mirrors and spatial filter setups (spatial filter and microscope objective) to the photo plate and the object. The scattered light from the object then reaches the photo plate representing the object wave. To improve the illumination it is favorable to use a second illumination wave which is drawn mirror symmetric to the first one in the figure. By this improved illumination unwanted shadows on parts of the object can be avoided.

All beams should be parallel to the table surface in order to be able to easily install all the components. The object and the photo plate should be hit as central as possible for optimal illumination. Therefore the beam adjustment is done without expanded beams at first. Instead of the photo plate a white cardboard is used to check the illumination during setup.

The expansion optics for the reference wave contains a spatial filter to cut out unwanted interferences. No spatial filter is required for the illumination wave if no too coarse interference structures are present (e.g., by dust particles on lenses).

Holography: A Practical Approach. Gerhard K. Ackermann and Jürgen Eichler
Copyright © 2007 WILEY-VCH Verlag GmbH & Co. KGaA, Weinheim
ISBN: 978-3-527-40663-0

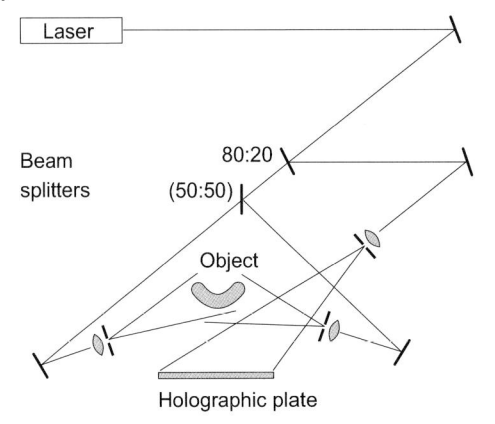

Fig. 10.1 Setup for producing a transmission hologram. The reference wave is guided through the beam splitter and by a mirror through the expansion system and onto the photo plate. The illumination wave is divided into two to achieve good illumination of the object.

10.1.2
Vibration

Unlike in single-beam setups the influence of vibration is more critical for two-beam holography. A vibration isolated table is therefore indispensable. The individual components need to be thoroughly fixed to the tabletop. Systems are common that use magnetic fixtures to hold all the components to the steel surface of the table. Less expensive is the use of hot glue with which all the components can be fixed to the table.

10.1.3
Object

Contrary to single-beam holography the distance of the object and the holographic film or plate can be large without surpassing the coherence length. Since reference and illumination beams are guided separately it is easy to keep the path difference smaller than the coherence length. Hence no principal restrictions are put on the distance of the object to the photo plate or the size of the object itself. However the depth of the object has to be smaller than half the coherence length. The hologram during reconstruction acts like a window for the observer through which he sees the image of the object.

The distance of the object from the photo plate is chosen as small as possible to receive a large viewing angle for the observer during reconstruction. For the same reason the object is always significantly smaller than the area of the photo plate. Today Slavich or Fuji plates or films are used mostly. Slavich films are delivered as 10 m up to 20 m rolls, the width varying between 3.6 cm and 1.2 m. Plate sizes range from 6.3 × 6.3 cm to 400 × 600 cm. For simple transmission holograms the object distance should be in the range of 10 cm whilst the object diameter should be around 5 cm.

10.1.4
Avoidance of Scattered Light

Before performing the final recording undesired reflections have to be avoided. The influence of the glass plates for holding the holographic film was already discussed in Chapter 9. Further scattered light is caused by every illuminated surface in the form of spherical waves. These cause in combination with the illumination or the reference wave unwanted interference patterns. Scattered light can be avoided by using screens (e.g., of black cardboard). Furthermore the reference wave should not hit the object and the illumination should hit only the object and not the holder. This can be wrapped in black cloth (e.g., velvet). Loosely fixed this cloth should suppress the light scattered by the holder without being holographically recorded itself.

10.1.5
Index Matching

The photo film is set up stabilized by glass plates. To avoid multireflections at the interfaces it is advisable to choose the Brewster angle between the reference wave and film. This quite large angle of about 56° can prove to be hindrance for some setups. In Chapters 8 and 9 it is described in detail how a liquid can be used to suppress the jumps in index of refraction glass/air/foil and therefore suppress the multireflections.

10.1.6
Visibility

The separate guiding of illumination and reference wave makes it possible to adjust their intensity ratio. The first beam splitter should divide the laser beam in a way that no more than 20% of the light is used for the reference wave. Since it impinges on the film directly after expansion its intensity is much higher than that of the light scattered by the object. Under some circumstances the reference wave has to be weakened further to achieve an acceptable visibility. The visibility is determined using Eq. (2.12a) from the measured intensities for the reference wave (I_R) and the object wave (I_O).

The intensity of the object wave is measured with a photometer at the brightest spots on the plate. With the setup in Fig. 10.1 it is quite possible to adjust a visibility close to $V = 1$. That the values in holographic practice are significantly below $V = 1$ is due to the H&D curve of the films (Fig. 13.1c). Very low and very high intensities cause strong deviations from the linear function. The difference between I_{max} and I_{min} must not exceed the linear region of the H&D curve. Therefore, it is favorable to select the intensity of the reference wave higher than that of the object wave. A typical value is $I_R \approx 10 \times I_O$. The resulting contrast is then $V = 0.57$. If the value for V approaches the maximum of 1 additional images of the first order and maybe other disturbing images appear due to nonlinearities.

10.1.7
Reconstruction of the Object Waves

The experiments on reconstruction that were presented in Chapter 9 can also be performed for two-beam holography. The reconstruction with the reference wave shows the orthoscopic, virtual image of the object. To get an impression of the diffraction efficiency, the hologram is placed into the original setup. By fading in and out the illumination wave the holographic image can be compared to the original object.

The reconstruction using the conjugated reference wave creates a real, pseudoscopic image that may be magnified under some circumstances. To do so the hologram is rotated by 180°, the axis of rotation being perpendicular to the table surface. So the direction of the reconstruction wave changes with respect to the hologram and the curvature of the reconstruction wavefront remains unchanged; only for plane waves this wave will represent the conjugated wave. For spherical waves the result is a magnified real image. A reconstruction with white light will give a spectrally blurred image (Chapter 9).

10.1.8
Reconstruction of the Reference Wave

To test the holography equation (2.22a) the hologram is illuminated with the object wave. For that the object is illuminated as during the recording and the reference wave is blocked off. With this method the reference wave can be reconstructed, although only weak. If parts of the object are covered these will be reconstructed by the remaining illuminated part of the object but very weak, too. Consequently parts of an extended object can always be an unwanted source of a reference wave if the visibility is high. Ghost images that are created this way lead to disturbing interferences with the actual image information. If the intensity of the reference wave is significantly higher during recording these effects do not play a major role. Therefore the visibility should not be chosen too high.

10.2
Setups for Reflection Holograms

Many topics discussed for transmission holograms, like contrast, unwanted reflections and questions of coherence length, also hold true for reflection holograms. The setup shown in Fig. 10.1 can be easily adapted to be used for making reflection holograms. In that case object and reference wave have to impinge on the hologram from opposite sides. The object is moved to the opposite side of the photo plate as shown in Fig. 10.2. The illumination waves are adjusted by the movable mirror and the expansion optics. The other components remain in position.

The path difference between object and reference wave is again measured from the first beam splitter to the photo plate to compare it to the coherence length. For most He–Ne lasers that are used in laboratories and that are operating at powers of 5 mW to 25 mW a path difference below 10 cm is uncritical since their coherence length is between 20 cm and 25 cm. Above this threshold the light loses its ability to interfere and no hologram is created.

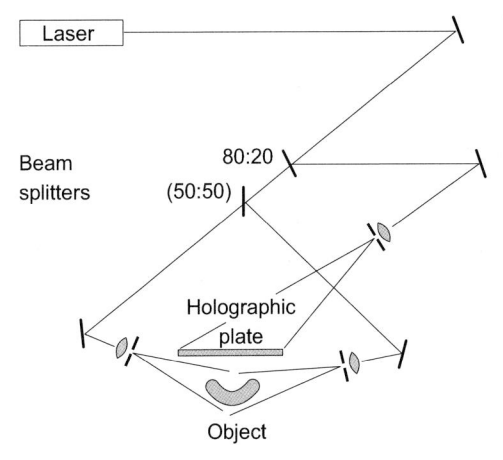

Fig. 10.2 Setup for creating a reflection hologram. The position of the object is mirrored with respect to the photo plate in Fig. 10.1. The illumination waves are re-adjusted accordingly.

Vibration-induced disturbances have a stronger effect in setups for reflection holograms than for transmission holograms. One fundamental source of disturbances is the movement of the photo plate. This becomes clear in Fig. 10.3. Vibrations perpendicular to the emulsion plane cause almost no changes in the interference pattern in the case of transmission holograms; however, for a reflection hologram the positions of the interference planes are shifted against each other and the recorded information is partly destroyed even if the amplitude of the motion is below $\lambda/2$.

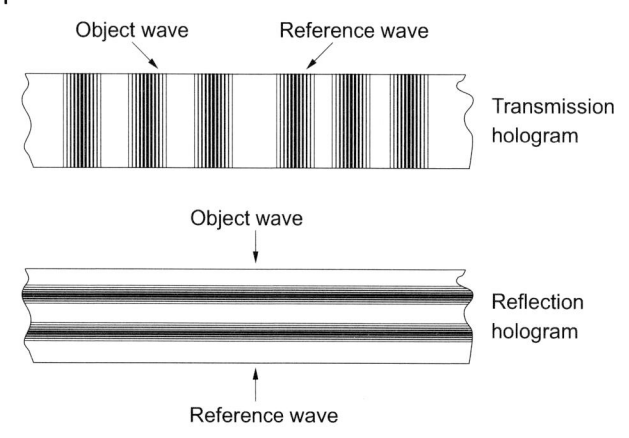

Fig. 10.3 Photo plate movement during exposure. Movement perpendicular to the surface destroys the interference pattern in the case of a reflection hologram. For a transmission hologram the movement is parallel to the interference pattern.

10.2.1
Object

The size and the positioning of the object depend on how the hologram is to be reconstructed. In most cases the hologram is illuminated with a light source at an angle of 45°, which is mounted at some distance above the observer. This way neither the light source nor the directly reflected light hampers the view of the reconstructed image. Thus the reference wave for an upright object has to come from above during recording.

For reasons of stability in most setups the reference wave is guided parallel to the table surface and the photo plate is illuminated from the side. According to this setup the object is rotated by 90° to make the described viewing possible. The above description shows that the details in the setup for the holographic recording are governed by the later reconstruction.

10.2.2
Reconstruction

For a reflection hologram the reconstruction is generally done with white light. The color of the image lies, depending on the developing process and the bleaching bath used, between red and green. It is governed by the distance of the grating planes in the emulsion. The image generation is only possible in a very tight wavelength region. Theoretically the diffraction efficiency drops to zero if the wavelengths differ more than 5% from the Bragg condition given by the laser and the developing process. Only this wavelength selectivity makes the illumination with white light possible. On the other hand, transmission holograms can only be reconstructed with monochromatic light, but with an arbitrary wavelength.

Problems

Problem 10.1 Assumed that the reflectivity of the object is 70%, calculate the visibility for the setup given in Fig. 10.1. The distance object/holographic plate is 15 cm, the plate is 10×10 cm.

Problem 10.2 Illuminating the hologram with the reference wave or with the object wave reconstructs the object wave and the reference wave, respectively. What is the reason for the always observed low intensity of the reconstructed reference wave (Figs. 10.1 and 10.2)?

Problem 10.3 If the area from the first beam splitter in Fig. 10.2 to the object is considered, the setup for the white light reflection hologram is a very symmetrical parallelogram. This is true for the left illumination wave and the reference wave. The right illumination wave is geometrical of the same length than the left one. What is the maximum possible distance between the object and the holographic plate? If the distance has to be larger than the maximum value, what measures can be taken to correct the setup in Fig. 10.2?

Problem 10.4 The white light reflection hologram (Fig. 10.2) is reconstructed with a white light point source at the same position than the pinhole of the reference wave ($z_r = z_c = 100$ cm). The distance of the object was $z_o = 15$ cm. The shrinkage of the emulsion is 20%. The reconstruction wavelength moved from 632.8 nm to 500 nm. What is the lateral magnification V_{lat} of the virtual and the real image? To view the real image, the hologram is rotated by 180°.

11
Experimental Setups for Holograms of Holographic Images

The selection of objects that can be recorded holographically is not limited to physical objects. Also the reconstructed real image of an object can be used as a holographic object. This way additional effects can be created. More complex methods that are presented in this chapter make it possible to let the reconstructed images float in front of the plate or protrude from it.

Rainbow holograms can create images in different spectral colors depending on the viewing angle. This type of hologram was developed by S. Benton in the late sixties. Meanwhile, a number of variants exist that in addition to the color effects simulate motion in the hologram (Chapter 12).

The general principle of such two-step methods was presented in Chapter 4. In the present chapter special technical advices on the creation of these hologram will be given.

11.1
Master Hologram (H1)

A master hologram, often called "H1," that is used as an object carrier for a secondary hologram (H2) is most often a transmission hologram (Figs. 10.1 and 11.1). Depending on the type of the secondary hologram some additional boundary conditions apply to H1.

11.1.1
Size and Position of the Object

In two-step holograms the source object for the creation of H2 is the real, pseudoscopic image of the object reconstructed by H1. The object itself should not exceed 5 cm in diameter for 9×12 cm^2 films and has to backscatter the laser radiation sufficiently. The reference is adjusted such that the reconstruction wave impinges "from above" onto the image for the observer of the reconstructed real image of H2. For stability reasons the reference wave is not adjusted "from above" during recording but from a more easily realizable angle from the side. Thus the object needs to be rotated by 90° into a sideward po-

Holography: A Practical Approach. Gerhard K. Ackermann and Jürgen Eichler
Copyright © 2007 WILEY-VCH Verlag GmbH & Co. KGaA, Weinheim
ISBN: 978-3-527-40663-0

sition during recording. When selecting the direction of the reference wave it has to be made sure that for both H1 and H2 the real image is reconstructed. Thus in both cases the reconstruction is done with the conjugated reference wave.

11.1.2
Object Distance and Position of the Reference Wave

The distance of the object to the master hologram is governed by two opposing requirements. If on one hand the object is placed closed to the plate the viewing angle, given by the object position and the size of the plate, gets very large and many observers can observe H2 simultaneously afterward since this angle is preserved during the transfer from H1 to H2.

On the other hand the zeroth diffraction order should not interfere with the area of the real image during reconstruction. Consequently, the reference beam has to impinge on the photo film at a flat angle during the above-described recording. This can lead to a significant decrease in diffraction efficiency (Chapter 6). Furthermore, the intensity recorded by the photosensitive layer is decreased. Thus the distance between the object and film must not be too small. One can see that it is favorable to use films as large as possible for master recordings so that the viewing angle during reconstruction does not get too small even for large distances of the object. The distance between the object and photo film should be 10 cm, for rainbow masters about twice this distance.

Figure 11.1 shows a setup that is very similar to the one in Fig. 10.1. The distance between the object and the plate is larger; the reference wave is plane in order to prevent magnification effects in the real image. If a spherical wave is used and the reconstruction of the real image is only done by turning the master plate the image is displayed magnified. This effect can be used specifically during the recording of H2.

11.1.3
Preparation of the Reference Wave

Before performing the recording the path difference is measured as was described in Chapter 10; unwanted reflections at parts of the setup are blocked by screens. Additionally, it has to be taken care of that no light reaches the film from the edges of the glass plates which hold the film (by multireflection). This scattered light leads to a very coarse interference pattern on the hologram (Section 9.4). In the end the visibility is measured and if necessary the reference wave is reduced if the visibility is too low.

According to the stability of the table the waiting period after loading the plate until exposure should be between 15 min and 30 min. For master holo-

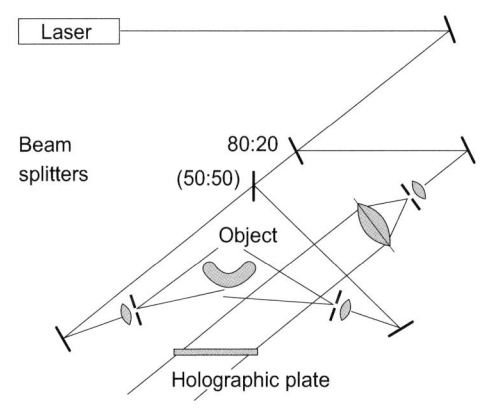

Fig. 11.1 Setup for a master hologram (H1). The first beam splitter has a low reflection efficiency in order to use as many light as possible for the object illumination. The reference wave is a plane wave. The expansion optics consists of a microscope objective (40×) and a pinhole (30 µm).

grams often plates are used, for larger masters also films due to costs. When using the "index matching" it is advisable to additionally tape down the edges of the film. It is recommended to use a vacuum film support (Section 9.1).

It is favorable to trigger the exposure from outside the laboratory and not to be present during the recording in order to avoid air movements. In some holography laboratories the finished setup is encased in a box or curtains are put around the table to keep air movements away from the recording area.

11.1.4
Duplicating Methods

Principally it is possible to make a contact copy of the master hologram. A film is brought in contact with the exposed and developed master in the setup of Fig. 11.1 and is only illuminated with the reference wave. The object wave is reconstructed by the illumination and leaves the master hologram together with the remaining reference wave. The two waves interfere with each other on the copy. This way a copy of the master hologram is created which can be used for further experiments instead of the original recording. In addition new copies of the master hologram can be created any time.

11.2
White Light Reflection Hologram (H2)

11.2.1
Single-Beam Method

In extension of the copying method Bjelkhagen [21] proposed a single-beam method for the creation of white light reflection copies (Fig. 11.2). He also uses a white light reflection hologram as a master as was described in Chapter 9 for single-beam setups.

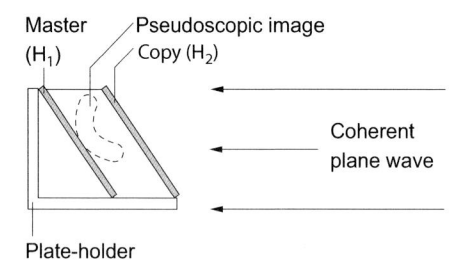

Fig. 11.2 Single-beam method for a white light reflection copy (H2) after Bjelkhagen [21].

In this method the reflection master hologram has to be bleached in rehalogenating bath so that H1 can be reconstructed with the recording laser. The tilt angle of the reference wave should be the Brewster angle in order to avoid multireflections.

11.2.2
Two-Beam Method

Figure 11.3 shows a setup for the production of a white light reflection hologram (H2); the principle is presented in Fig. 4.3. The master hologram is set up in a way that the real, pseudoscopic image appears in front of the plate. (It would also be possible to use the virtual orthoscopic image for the production of an H2. However this would lie behind the plate and the white light reflection hologram would again create an image behind the plate because only this one is also orthoscopic.) The real image allows manipulations during the setup that would not be possible with a real object. The position of the white light copy can be chosen such that the object lies partly behind (virtual) and partly in front (real) of the plate or completely in front or behind the plate.

After exposure and developing the H2 hologram the pseudoscopic, real image of the used H1 image can be reconstructed. But since this was also pseudoscopic the observed real image of H2 is orthoscopic.

After finishing the setup for H2 the path differences and the visibility are checked. The intensity of the object wave is measured at the brightest spots

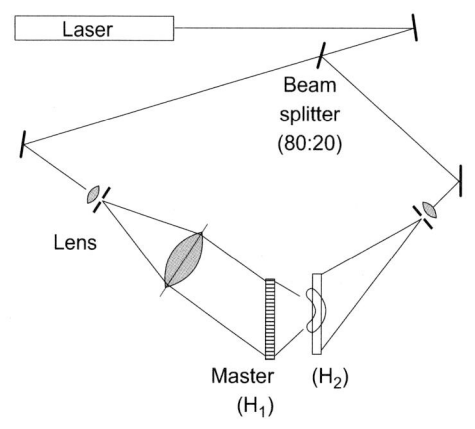

Fig. 11.3 White light reflection hologram (H2) of a master according to Fig. 11.1. The reference wave is oriented such that the reconstruction wave for the real image of the H2 impinges from above.

of the reconstructed image (H1) at the position of the H2 recording. Problems with intensity concerning the object wave are smaller for the H2 recording since the whole intensity of the reconstructed image can be used and no restrictions due to weakly scattering object areas apply.

Principal disadvantage of an H2 hologram is that it can only be viewed in the angular area given by the size of H1 and the object distance. Only that area of H2 carries image information that is directly hit by the reconstructed image waves of H1. The rest of H2 does not contribute to image creation during reconstruction. Therefore the size of H1 should be selected as large as possible when planning such recordings while H2 does not need to be larger than the object at the recording position.

11.2.3
Image Aberrations

The hologram acts like a complicated Fresnel lens. Therefore the reconstruction shows all the image aberrations known from optics: spherical aberration, coma, astigmatism and chromatic aberration to name the most important aberrations.

Benton put together the basic equations for the calculation of image aberrations [22] and evaluated them to extend for the case of rainbow holography. Hariharan [3] did put these formulas in general from. In the end these aberrations lead to the fact that the reconstruction of the image recorded in H1 is out of focus. Most of the time the spherical aberration is the largest and most serious geometrical error. For a lens it increases very strongly from the center to the edge. Therefore, the edge areas are blocked off by apertures in imaging processes using lenses. However, for H1 such a measure would further limit the viewing angle to watch the image of H2.

Coma and astigmatism increase with the angle of the individual waves to the optical axis. Consequently, they do not occur if we reconstruct with beams parallel to the optical axis and use parallel reference waves. This can often not be managed due to the use of large optical components which are needed for the illumination of large hologram plates with a plane wave. In different setups a divergent beam is used for the reconstruction of the real image although the required conjugated wave should be convergent. Thus the distance between the hologram and real image and thus the expected image aberrations are increased. Therefore, blurred areas due to aberrations have to be expected for a lot of recordings especially if the reconstructed image is far off the front of the plate. If the recording and reconstruction geometry do not differ no image aberrations occur.

11.3
White Light Transmission Hologram (H2)

11.3.1
Image Plane Hologram

Transmission holograms unlike reflection holograms have the advantage of a higher brightness. The realization of a so-called image plane hologram is shown in Fig. 11.4. The reconstruction wave for the master hologram (H1) is guided divergent to H1. Depending on the recording technique for the creation of the master (H1) this reconstruction wave can also be collimated.

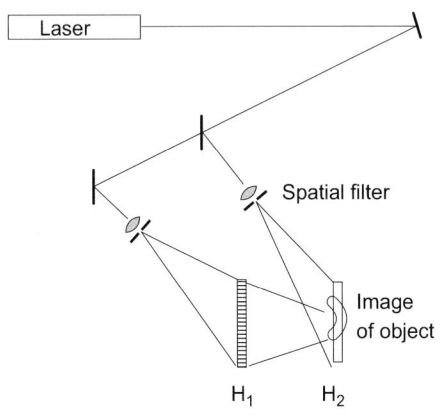

Fig. 11.4 Setup for a white light transmission hologram. For an image plane hologram the photo plate for H2 is placed in the image plane of the real image of the master hologram (H1).

The reference wave impinges on the photo film for H2 from the same side as the reconstruction wave: accordingly a transmission hologram is created. The plate H2 is positioned in the image plane of the reconstructed object; an "image plane hologram" is created (Section 4.3). This hologram is called a "full aperture hologram" in contrast to the "slit aperture" hologram following below.

Although H2 is a transmission hologram for which monochromatic light would be needed for reconstruction this hologram can be reconstructed using incoherent white light. Since the image is reconstructed in the plane of the hologram the linear dispersion is still low despite the fact that an angular dispersion is present; the individual spectral colors are not noticeably separated in space. If the image plane of H1 is not used for the creation of H2 it shows again during reconstruction with a white light source that every hologram is a kind of diffraction grating. The image appears in all colors simultaneously but spatially separated as in a continuous spectrum.

11.3.2
Rainbow Hologram

Rainbow holograms or "slit aperture" holograms are a variation of the just described arrangement. They were invented by S.A. Benton in 1969 and are characterized by a particular richness of color. The effect that the object reconstructed by H2 can be observed in the colors of the spectrum ("rainbow") is created by the fact that only a stripe of H1 and not the complete master hologram is used for the production of H2. The basics of this technique are explained in Section 4.3.

First of all a master hologram is created as shown in Fig. 11.1. For the reconstruction only a part of the master hologram is used which is cut out with a small slit of about 5 mm width (Fig. 4.4). If only rainbow holograms are to be created from the master it is possible to use only stripe of film with dimensions of the slit. Optically a cylinder lens can be used for the illumination of a small stripe of H1 for reconstruction. The advantage of this method in contrast to using a geometric slit is the increased intensity of the object wave. The slit in Fig. 11.5 is perpendicular to the plane of the paper. The geometric edge of the slit is recorded in H2.

The slit image lies between H2 and the observer when reconstructing with the conjugated beam and is oriented horizontally as the human eyes (Fig. 11.6). The perspective and the three-dimensional vision for the observer are not limited in the horizontal direction by the slit; however in the vertical direction the three-dimensional vision is lost. Yet this is not noticed by the observer.

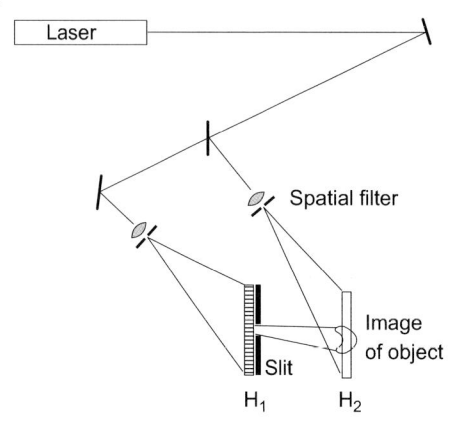

Fig. 11.5 Setup for a rainbow hologram. Additionally to the reconstructed image of the object an image of the slit aperture, perpendicular to the plane of paper, is recorded in H2.

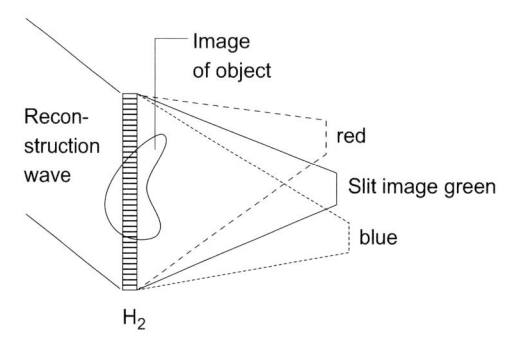

Fig. 11.6 Image reconstruction for rainbow holograms. Reconstructed are the images of the object and of the slit. The image of the object is reconstructed close to the H2 (image plane hologram) while the slit is reconstructed in all colors at a larger distance to the hologram plane.

11.3.3
Reconstruction

During the display of the image, e.g., with the light of the used laser, the image of the object is reconstructed (Fig. 11.6). The image can only be observed though the slit as was described shortly in Section 4.3. The image lies close to the hologram plane just like for an image plane hologram; the slit is at the distance of H1 from H2. At first no advantage of this technique against the before described technique is evident. However, if instead of the recording wavelength λ_0 another wavelength λ_1 is used for reconstruction the object is still reconstructed. Although the corresponding slit image lies higher than the one for λ_0 if $\lambda_1 > \lambda_0$, it lies lower if $\lambda_1 < \lambda_0$. The hologram is a diffraction grating and delivers for short wavelengths ("blue") a small, and for large wavelengths ("red") a large diffraction angle.

The image is reconstructed quite sharp when using white, incoherent light as is the case for all image plane holograms. The slit which does not lie in the image plane is reconstructed in different colors and at different diffraction angles depending on the color. The slit images therefore lie in front of the hologram above each other and sorted by color.

The size of the slit can be calculated from diffraction theory. It should not be too large; otherwise the spectral colors cannot be separated anymore. If it is too small the speckle structure of the laser becomes a disturbing factor. A slit width of about 5 mm is acceptable for many purposes.

A special effect is created if the hologram is illuminated from different angles simultaneously such that large parts of the spectrum are reconstructed at the same place. Then a more or less achromatic image becomes visible. To do this a vertically extending reconstruction light source as it is represented by a fluorescent tube is needed. With such a spatially extended light source the image appears black and white (achromatic).

11.3.4
Calculation of a Rainbow Hologram

The production of a rainbow hologram is not more difficult than other transmission holograms. By preceding calculations it can be set at which angle and at which distance the spectrum appears. These questions are especially important for applications of rainbow holography in arts.

11.3.5
Optical Basics

If a hologram with the grating constant d_g is illuminated at an angle α and the first diffraction order is observed at an angle β, then

$$d_g = \frac{\lambda}{\sin\alpha + \sin\beta}. \tag{2.35}$$

The connection between the image distance s_i, the object distance s_o, and the focal length f of a lens is given by the imaging equation:

$$\frac{1}{s_o} + \frac{1}{s_i} = \frac{1}{f}. \tag{11.1}$$

This imaging equation is interpreted a little bit different than it is usually done in geometrical optics. Especially it has to be taken into account that the chromatic aberration in holograms is very large. The focal length is a function of the wavelength. Equation (11.1) is adapted to the problem of the "reconstruction" of a slit by a reconstruction light source that illuminates a hologram. The distance of the reconstruction light source is denoted with r while the distance of the real image is called s:

$$\frac{1}{r} + \frac{1}{s} = \frac{\mu}{f}. \tag{11.2}$$

In this equation μ is the ratio of the wavelengths during reconstruction and recording and f is the focal length of the hologram for the real image. Equation (11.2) cannot be directly compared to Eq. (5.3), where the z-components of the real image and the virtual image are connected with the focal length. With the help of Eq. (11.2) it is possible to calculate the distances of the slit images from the hologram that appear in a rainbow hologram for the red, green and blue region. In the literature the recording wavelength λ_r and the focal length are united in a product f^* that is set during the recording. It is more clear to use the parameter μ from Chapter 5.

11.3.6
Calculation Example

As an example it is assumed that the green light ($\lambda = 550$ nm) of a continuous spectrum of a light source at infinite distance, e.g., the sun, shall reconstruct the slit image at a distance of 45 cm from the hologram. The light is impinging at an angle of $45°$. According to Eq. (11.2) f is given by the above chosen distance of the slit image and with the according μ the red slit image ($\lambda = 630$ nm) lies at 39 cm whilst the blue one ($\lambda = 450$ nm) lies at 55 cm.

Equation (2.35) is used to calculate the angles at which the slit images in the different colors are observed. The assumption being made is that the green slit image emerges perpendicular from the plate ($\beta = 0$) if the hologram plate is illuminated at $45°$. The result is $d_g = 780$ nm. For the above chosen wavelengths the calculated diffraction angle is $+5.8°$ (red) and $-7.4°$ (blue), respectively. Prerequisites for the observation of the finished rainbow hologram set the experimental conditions which have to be obeyed when creating H2 and H1.

11.3.7
Designing a Rainbow Hologram

For the design of a rainbow hologram the grating equation (2.35) and the image equation (11.2) are used. This procedure works for thin holograms only. But also for volume holograms the results are approximately right. The design is planned from the end result where it is decided which color appears at which angle with a given direction of the reconstruction wave. That way the parameters of the rainbow hologram (H2) are set which again set the recording geometry of the master hologram (H1).

The grating constant of H2 is given by the direction of the reconstruction wave and the desired observation direction (position of the slit image) for the reconstructed image of the object in Eq. (2.35). The focal length f is given by setting the distances between the slit image and H2, s, during the reconstruction and the distance of the light source, r, and the parameter μ. All parameters are given by the assumption. For the calculation Eq. (11.2) is used. With the so calculated f the resulting distance s_o of the slit from H2 during the recording of H2 can be calculated from Eq. (11.2) in the form ($\mu = 1$)

$$\frac{1}{s_o} - \frac{1}{r} = \frac{1}{f} \tag{11.2a}$$

If it is not possible to use a plane reference wave the distance of the reference light source should be as large as possible.

The focal length of the master hologram is calculated from Eq. (11.2a) with the just acquired value for s_o and the distance of the reference wave for $\mu = 1$ being as large as possible. The distance s_o represents the distance of the object image from the master hologram during reconstruction (H2 is an image plane hologram). The distance of the object during recording of H1 is given by Eq. (11.2a). Equations (11.2) and (11.2a) differ by a sign because in the first case the distances r and s are on the opposite sides of the hologram whilst in the case of Eq. (11.2a) they are not.

Problems

Problem 11.1 Show that reconstructing the virtual or the real image with the reference wave of the same wavelength or its conjugated complex, the lateral magnification is 1 (Fig. 11.1).

Problem 11.2 In Fig. (11.3) the setup for a white light reflection hologram (H2) is given. For the ratio of the intensities of object wave I_o and reference wave I_r it should be $(I_o/I_r) \leq 1$. What is the maximum diffraction efficiency of H1 to fulfill this condition? Change the setup for a diffraction efficiency of 45% in order to fulfill the condition given above.

Problem 11.3 Calculate the beam splitter in Fig. 11.4 assumed that the diffraction efficiency is 50% and the visibility should not exceed $V = 80\%$. The image size of the image is 5×5 cm^2 and of the holographic plate 10×10 cm^2.

Problem 11.4 Calculate the beam splitter for Fig. 11.5; assuming the same data given in Problem 11.3. The slit is about 5 mm wide and 10 cm long. The plate is again 10×10 cm^2.

Problem 11.5 Design a rainbow hologram, using a He–Ne laser ($\lambda = 632.8$ nm) for the master and the H2 hologram. The blue image of the object shall be reconstructed under a diffraction angle $\alpha = 0$ (perpendicular to the H2 surface). The reference wave and the surface of H2 shall include an angle of 60°. The distance of the blue image of the slit from the rainbow hologram shall be 30 cm. Calculate the slit image for green ($\lambda = 500$ nm) and red ($\lambda = 600$ nm) light too. The reconstructing white light point source is very distant ($r \Rightarrow \infty$).

12
Other Methods in Holography

The so far described basic methods for the production of transmission and reflection holograms have been extended by the application of additional effects. From the abundance of proposals a few important will be presented here. New developments are found in [23]. In order to follow the developments in holography, the results of the upcoming SPIE publications should be observed.

12.1
Shadow Hologram

The holographic display of transparent objects with the so far described methods is difficult because the reflection efficiency of such objects is low. Either the visibility is very low or the reference wave has to be decreased a lot. However, these measures increase the exposure time and therefore the influence of disturbing effects, e.g., vibration.

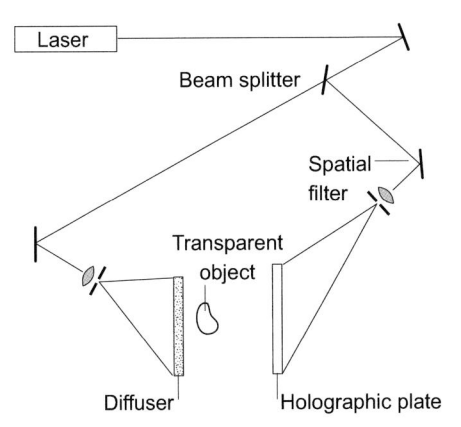

Fig. 12.1 Setup for a shadow hologram. The transparent object is illuminated diffuse. This way a shadow of the object is created.

In a shadow hologram the high transmission ability of the object is used. The object wave is guided through a diffuser (frosted glass) which needs to be

Holography: A Practical Approach. Gerhard K. Ackermann and Jürgen Eichler
Copyright © 2007 WILEY-VCH Verlag GmbH & Co. KGaA, Weinheim
ISBN: 978-3-527-40663-0

mounted vibration safe and passes through the transparent object. A setup for a white light reflection hologram is shown in Fig. 12.1. If the object of choice is a photo slide or a flat object from which only the contours are needed both the real and the virtual images can be used for reconstruction because the pseudoscopy of the real image is not observable.

This setup can be used to make a hologram of a real shadow, using an opaque object. This kind of shadowgram was popularized by Rick Silberman in 1979 during an exhibition of his work in the New York Museum of Holography. The museum was closed in the meantime.

12.2
Single-Beam Rainbow Hologram

There are many ways to produce a rainbow hologram without a master hologram using single-step methods. Here the object is imaged 1:1 in the plane of the film using a lens. The slit is positioned between the object and the lens. The conjugated reference wave is used as the reconstruction wave. Due to the imaging process by the lens the reconstructed image is orthoscopic. Besides the object the different slit images are reconstructed through which the object can be seen in the colors of the spectrum. One disadvantage of this setup is that a lens with large diameter is needed which also limits the field of view.

In an easy method for the production of a rainbow hologram the just mentioned shadow hologram setup is used. In a Dan–Schweizer hologram [24] the object wave that is used for the illumination of the slit is guided through a diffuser just like in a shadow hologram setup. The diffuse slit light source irradiates a transparent object that is positioned close to the photo plate. The small distance is necessary for rainbow holograms to minimize the linear dispersion. If the object is too far away from the plate during the recording the reconstructed image is blurred (Chapter 11).

Figure 12.2 shows a setup for the production of a Dan–Schweizer hologram. The conjugated reference wave is used for reconstruction. Instead of placing the object close to the plate a flat object (slide, silhouette) can be brought in direct contact with the photo plate. Although the image, as well as the object, is not three-dimensional it appears on a color changing background. Often this method is combined with other effects, e.g., multiple exposures, whose results will be described in the next section.

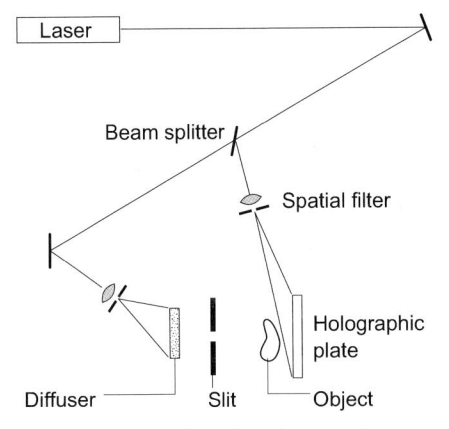

Fig. 12.2 Production of a rainbow hologram without master (Dan–Schweizer hologram). To create the shadow a slit light source is used.

12.3 Multiple Exposure

After the holographic recording of an object and its storage on the holographic film there are still enough unexposed silver bromide particles for another hologram inside the emulsion. The interference patterns of two recordings are totally independent of each other. Multiple exposures of the same object play an important role in interferometry (Chapter 15).

In this section methods are presented that are either not realizable or not sensible in photography. At first it is shown how different scenes or objects can be holographically recorded on the same plate by multiple exposure. In another technique the object is recorded several times next to each other on individual parts of the plate while the remaining unexposed part is blocked off. By this method one could also create a photographic recording but this would not have an advantage compared to multiple recordings which are done the usual way after each other. In the multiple exposure method described in the following the plate is not developed between the exposures.

If the images of two objects are to be reconstructed at the same place in space penetrating each other the object needs to be changed in the darkroom after the first exposure. The second recording is done with the same setup. Instead of two objects two master holograms can also be used; if the two masters can be recorded using the same geometry the method can be reduced to a single step in which both reconstructed images are recorded simultaneously.

Using multiple recordings also motions in the hologram can be simulated or an abrupt change of scene can be suggested. Several scenes are recorded in one hologram, e.g., by using different directions of the reference beam. Rotating then the finished hologram the reconstruction wave displays all images after each other.

On the other hand using a fixed reference wave several objects or the same object in different positions can be recorded in one hologram. This is done, e.g., for two recordings, by blocking off one half of the hologram and recording the first setting. Afterward the other half is blocked off and the second recording is done with the changed setting. If the finished hologram is illuminated with the reference light source the observer sees the image changing when walking by. This gives the impression of a motion.

Multiple exposures on the same film can only be done within certain limits because the H&D curve is not linear. There may be not enough unexposed AgBr particles anymore; this leads to a loss in modulation and therefore decreases the diffraction efficiency.

12.4
Multiplex Holograms

In multiplex holograms the method of exposing individual parts of the photo film and covering the rest is consequently used to simulate motion in the image. A multiplex hologram is based on the well-known principle of stereoscopy. In this method two photographic images of a scene are taken from two different perspectives. Viewed through a stereoscope the imaged objects appear three-dimensional for the observer.

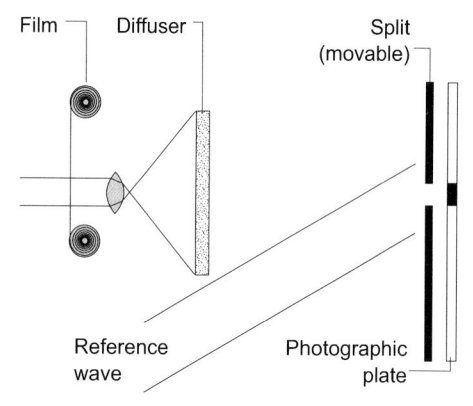

Fig. 12.3 Setup for the recording of a multiplex hologram. An incoherent recording of a scene consisting of many film images is imaged picture by picture as a stripe hologram.

Translated this idea to holography, a changing arrangement of objects can be recorded with a usual film camera and then exposed image by image onto a holographic film (see Fig. 12.3). During reconstruction different two-dimensional image information meet the left and the right eye of the observer. Without any other viewing aid the images are seen three-dimensional by the observer just like in a stereogram. When walking by he perceives the sequence of images as motion.

12.5
360° Holography

The 360°-holography is supposed to make the complete geometry of the object viewable. Figure 12.4 shows the setup for a single-beam transmission hologram. When illuminating the finished hologram with a laser all perspectives of the object are displayed when walking around the hologram.

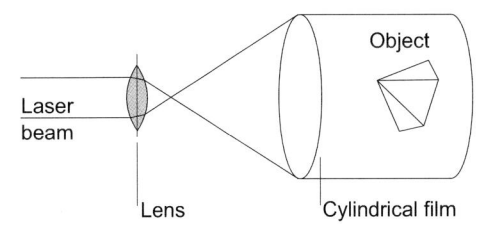

Fig. 12.4 Setup for a single-beam recording for a 360° transmission hologram.

Multiplex holograms (whose recording technique was described above) are also often reconstructed using this 360° geometry. With an illumination from above all recordings of the multiplex hologram can be reconstructed simultaneously and the recorded changing images can be observed when walking around the complete circle. This way the 360° technique only serves the purpose of a better presentation. Additionally the image reconstruction is easier this way. For a longer sequence consisting of many individual images arranged in a linear way it would merely be possible to illuminate it with only one light source obeying all the necessary reconstruction angles.

12.6
Color Holography

All holograms deliver colored but mostly monochromatic images. One exception is the achromatic reconstruction which leads to a black and white image. There are also several different possibilities of changing the color set by the wavelength of the laser used or to use the complete spectrum by using a rainbow hologram and reconstruction with white light. These pseudocolors that do not resemble the real colors of the object are known in holography as well as in photography. False colors are used in photography, e.g., in satellite images to display the different reflection ability of land formations in the infrared. In this section holographic techniques to display true colors are presented.

12.6.1
Multilaser Techniques for Transmission Holograms

For true color holographic images three colors, blue, green, and red, are needed to resemble the spectrum reflected by an object realistically. To do so a blue and a green line of an argon-ion laser can be used besides the red line of a He–Ne laser. A principal setup for a transmission hologram is shown in Fig. 12.5. The first beam splitter reflects light with short wavelengths and transmits light with long wavelengths.

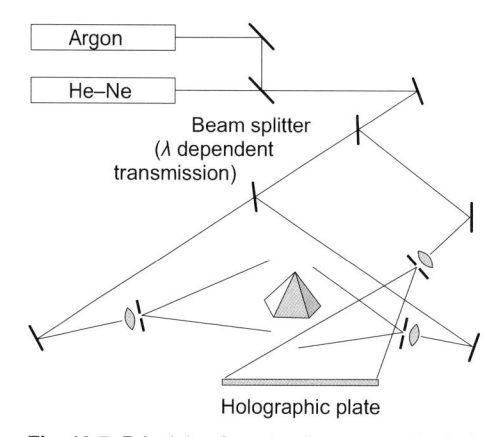

Fig. 12.5 Principle of a setup for a true color hologram. The first beam splitter reflects light with short wavelengths and transmits light with long wavelengths. The setup allows it to create the true color hologram in one recording.

The problem with transmission holograms is their low wavelength selectivity. Each of the three holographic recordings is able to reconstruct all three colors; therefore not only three but nine images are reconstructed although not with equal intensity. These partly superpose the true color image and lead to a decrease of the diffraction efficiency and to an unwanted blur. This problem can be avoided by using volume holograms. The Bragg condition only allows the reconstruction with the recording wavelength. All other directions or wavelengths violate this condition and are strongly suppressed. The object is displayed in its true colors when reconstructed using the recording lasers. However, this procedure is very expensive.

12.6.2
Multilaser Techniques for Reflection Holograms

Volume holograms have a large wavelength selectivity. They reconstruct each of the three different colored images in only one color and it is possible to use white light for reconstruction. A single-beam setup is shown in Fig. 12.6. The problem of true color holography is divided into two steps if no panchromatic emulsion is available.

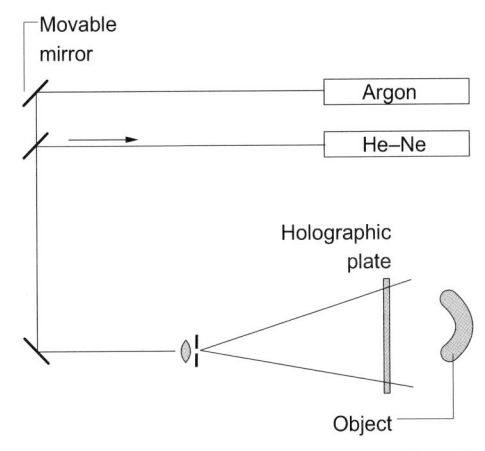

Fig. 12.6 Single-beam setup for a true-color reflection hologram. One mirror is movable for sequential exposure with the He–Ne and the Ar lasers.

Using the setup in Fig. 12.6 two images are exposed on two spectrally different sensitive films; therefore one mirror is movable in the indicated direction. Both plates are tightly united after the developing. If plates are used it has to be made sure during recording already that photo layer is on top of photo layer during reconstruction because only then different monochromatic images can appear at the same spot.

Although reflection holograms are quite color selective due to the Bragg condition when reconstructed with white light the result is not truly a superposition of the monochromatic images that were recorded with lasers. Jeong and Wesly [25] proposed a definition for a true-color hologram almost 20 years ago. According to their definition a true color hologram is only given when it reconstructs the same wavelengths with the same relative intensities that were reflected from the object during the recording. This condition cannot be fulfilled with the above method. The image reconstruction for reflection holograms is often done with a different wavelength than the one that was used for recording. In a solving bleaching bath the developed silver is removed from the emulsion, the photo sensitive layer shrinks, and the reconstruction wavelength decreases. The colors can be shifted from "red" to "green" for the red wavelength and from "green" to "blue/violet" for the green wavelength. This way a color hologram with true colors is barely realizable. With rehalogenating bleaching baths and developers which only cause minor shrinkage this shift can be greatly reduced.

Triethanolamin of different concentration can be used to soak the emulsion layer and to achieve a specific shift in wavelength after exposing and developing. Besides, from the problem of creating a true color hologram this way the holograms still darken with time due to the creation of photolytic silver. Hariharan [3] proposes to use D-sorbitol instead of triethanolamin ($CH_2OH(CHOH)_4CHOH$). The well-aimed wavelength shift by chemical treatment is mostly used to create false-color holograms.

12.6.3
Color Holograms Using Rainbow Technique

Another method for the creation of true color holograms uses the rainbow hologram technique [26, 27]. Color shifts due to shrinkage cannot happen because these are transmission holograms. Three master holograms are produced in the red, green, and blue colors. The masters are chosen in a way that during reconstruction of the red image the red slit image, for the green image the green slit image, etc., have the same impinging angle with respect to H2. During the reconstruction of the H2 with white light (see Fig. 12.7) three different arrangements of slit images are created. In a certain direction from H2 the three desired colors superpose and form a true color hologram. If it is observed more from the top or the bottom first a false color hologram is reconstructed and in the end only a monochromatic one either red when viewed from above or blue when viewed from the bottom.

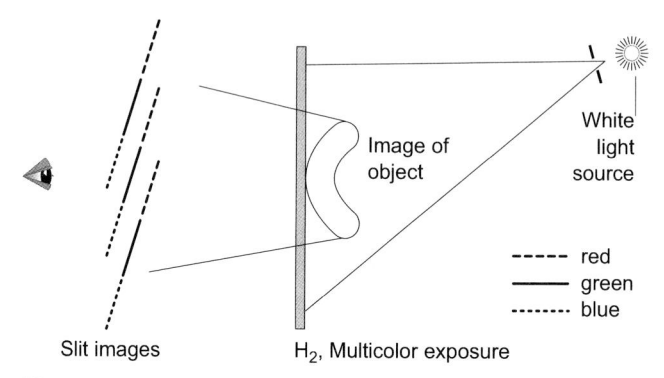

Fig. 12.7 True color hologram using the technique for a rainbow hologram. Shown is the reconstruction of images of a hologram that was created by the superposition of three master holograms in red, green, and blue. In a tight area the slit images of all three colors superpose; the observer sees the object in its natural colors.

12.6.4
Achromatic Images

In Section 11.3 the achromatic reconstruction of rainbow holograms was mentioned. According to the method described there the slit images for the three slits that were recorded, e.g., with a He–Ne laser can be calculated such that in the reconstructed H2 red, green, and blue part from one slit at a time partly overlap. In this case if white light is used for illumination the image of the object appears achromatic, i.e., black and white, in the overlap region.

Problems

Problem 12.1 In Fig. 12.2 a setup for a Dan–Schweizer hologram is shown. It is assumed that the beam splitter data are 4:1 (transmission: reflection). The He–Ne laser ($\lambda = 632.8$ nm) has an output of 25 mW. The slit in front of the diffuser shall get 30% of the illumination beam. The slit geometry is 5 mm × 10 cm, and the distance of the slit from the photo plate (size 10 × 10 cm^2) is 15 cm. The transparency of the object shall be close to 100%. Calculate the visibility and the exposure time, if the sensitivity of the emulsion PFG-03M is 1500 µJ/cm^2 (Table 14.2).

Problem 12.2 In Fig. 12.4 the setup of a 360° transmission hologram is given. The object shall be a cone with a base of 10 cm in diameter. What is the maximum diameter of the photographic cylinder? What happens if the diameter is slightly larger than the maximum value?

Problem 12.3 In Figs. 12.5 and 12.6 two setups are given to make real color holograms. What is the main problem of making color holograms (see Fig. 12.7)?

Problem 12.4 In Fig. 12.7 the reconstruction of a real color hologram is presented. The wavelengths of the three master holograms were 488 nm, 514 nm, and 632.8 nm, respectively. Calculate the positions of the six other images next to the real color image of the image plane H2 hologram. The directions of reference waves as well as reconstruction white light point source were 45° with respect to the surface of H1 and H2, respectively.

13
Properties of Holographic Emulsions

Different recording materials are available for holography; this chapter contains a general characterization of holographic emulsion layers. Special holographic media are the topic of this chapter.

13.1
Transmission and Phase Curves

The optical properties of holographic emulsions change during exposure and subsequent chemical development. Through this the amplitude and the phase of a light wave passing the emulsion are influenced. The equation for the change of the complex amplitude transmission by the exposure is

$$t = e^{-\Delta\alpha d} e^{i\Delta\Phi} = t \cdot e^{i\Delta\Phi}. \tag{13.1a}$$

The modification of the absorption coefficient caused by the exposure is given by $\Delta\alpha$. The index of refraction n and the layer thickness d change by Δn and Δd, respectively. The phase is shifted by

$$\Delta\Phi = \frac{2\pi\Delta(n \cdot d)}{\lambda} = 2\pi\frac{\Delta n \cdot d + n \cdot \Delta d}{\lambda} \tag{13.1b}$$

with λ being the wavelength. Depending on the emulsion and the development process amplitude holograms can be created if the exposure only changes the absorption coefficient α, or phase holograms if either the refraction index n or the layer thickness d is changed (Section 6.2). The sensitivity of the holographic material in these two borderline cases is given by the amplitude and phase curves shown in Figs. 13.1a and b which describe the dependence of these parameters on the exposure E. The exposure E is calculated from laser power $P \cdot$ exposure time τ; for pulse lasers E is the pulse energy.

13.1.1
Optical Density

Especially for holographic layers it is more common to use the term "optical density" instead of the amplitude transmission t. It can be calculated from the

Holography: A Practical Approach. Gerhard K. Ackermann and Jürgen Eichler
Copyright © 2007 WILEY-VCH Verlag GmbH & Co. KGaA, Weinheim
ISBN: 978-3-527-40663-0

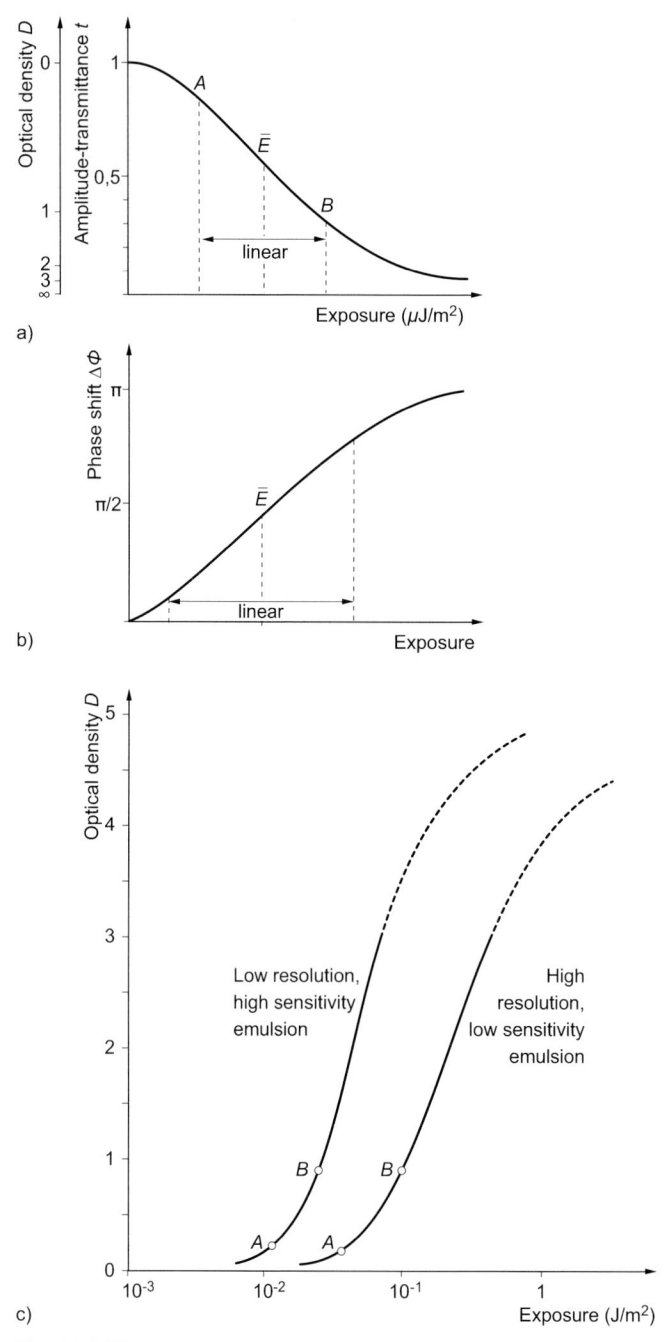

Fig. 13.1 Transparency and phase curves of holographic materials [3, 8]: (a) amplitude transparency t of a holographic emulsion layer as a function of the light energy E, (b) shift of the phase $\Delta\Phi$ as a function of the light energy E, and (c) optical density $D = \log 1/t^2$ as a function of E (Hurler and Driffield, H&D curve).

intensity transmission $T = t^2$ according to

$$D = \log \frac{1}{T} = \log \frac{1}{t^2} \quad \text{or} \quad D = -\log T = -2\log t. \tag{13.2}$$

In photography it is common to show D as a function of $\log E$ (Fig. 13.1c).

13.1.2
Modulation

In holography usually a linear recording characteristic is desirable. This means that the transmission t of the layer is proportional to the incident light energy E $(= I\tau)$, see Eq. 2.10. In this case sinusoidal grating structures are created and only the 0th and the $\pm$1st diffraction orders appear.

To investigate if a linear recording is possible, e.g., a light grating with a brightness distribution of $E \propto \sin(x)$ is observed (more exactly, $E = \bar{E} + \bar{E}\sin(kx)$, x = spatial coordinate, $k = 2\pi\sigma$, σ = spatial frequency = 1/grating constant, $\bar{E}$ = mean light energy). This grating is to be recorded in a holographic layer, i.e., is to be photographed. It can be seen from Fig. 13.2a that the resulting transparency is not at all a linear function of E. The recorded grating shows strong deviations from the $\sin(x)$ function; distortions, i.e., higher harmonics, appear which cause higher diffraction orders. In holography this is unfavorable since during reconstruction multiple images at different angles can appear.

Figure 13.2b illustrates that it is nonetheless possible to use the almost linear part of the t–E curve. To do so a constant background, i.e., a homogeneous illumination, is added: $E = \bar{E} + \bar{E}K\sin(kx)$. The visibility V (see Eq. (2.12a)) of the grating is smaller than 1 such that the intensity minima are no longer zero. Thus the light grating is shifted into the linear part of the curve and the transparency despite a constant background also becomes a $\sin(x)$ function. The background does not affect the diffraction so that this method of linear recording only yields the 0th and $\pm$1st diffraction orders.

13.1.3
Bleaching

For amplitude holograms an almost linear recording with high amplitude according to Fig. 13.2b is relatively easy to achieve. This process is much more complicated for phase holograms which are created by bleaching of the emulsion layers. In Section 6.2 it was described that for thin phase holograms always several diffraction orders appear. This is also true if the phase shift $\Delta\Phi$ grows linearly with the exposure E. Only for thick phase holograms due to the Bragg effect only one diffraction order appears.

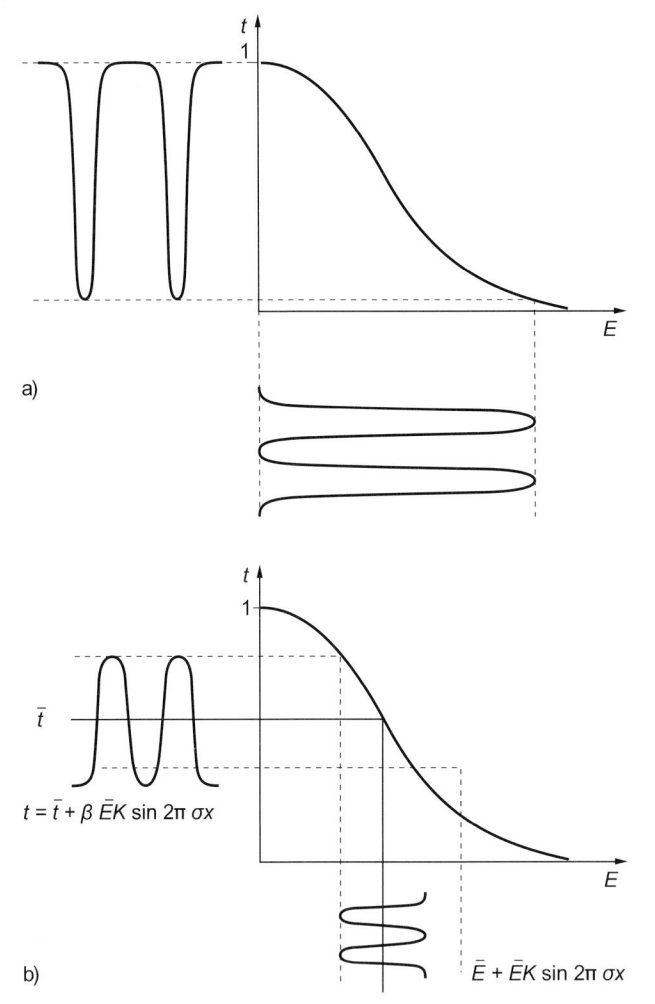

Fig. 13.2 Recording of gratings in holographic layers [27]: (a) For strong changes of the intensity the recording is nonlinear. A distorted $\sin x$ grating is recorded. (b) Linear recording is achieved for small grating amplitudes if a constant light background $\bar{E}$ is superposed, i.e., the contrast has to be less than 1.

13.2
Resolution and Diffraction Efficiency

13.2.1
Visibility Transfer Function

The curves of the transmission and the phase in Figs. 13.1 and 13.2 represent the macroscopic behavior of recording materials. Information on the resolution is contained in the visibility transfer function which depends on the spatial frequency σ during the recording. It is defined by the visibility V' after recording a grating with the contrast K (Eq. (2.12a)):

$$M = \frac{V'}{V}. \tag{13.3}$$

13.2.2
Recording

In the following the recording of a grating in a holographic layer is calculated using Fig. 13.2b. The distribution of the brightness is given by

$$E = \overline{E} + \overline{E}V \sin kx. \tag{13.4}$$

Here $\overline{E}$ is the mean light energy and x the spatial coordinate; $k = 2\pi\sigma$ is governed by the spatial frequency σ. This light grating leads to a transmission of the layer of

$$t = \overline{t} + \beta\overline{E}MV \sin kx, \tag{13.5}$$

where the visibility transfer function was taken into consideration. The rise of the t–E-curve in the linear part is denoted by β ($= dt/dE$). The equation is only valid in the linear region of the recording; of course $t < 1$ and $\beta < 0$.

13.2.3
Efficiency

The amplitude of the wave diffracted by the grating can be read from Eq. (13.5) if the grating amplitude $\beta\overline{E}MV$ is multiplied by the illumination wave, which is identical to the reference wave r. For the amplitude u_3 of a holographic image obeying the factor of 0.25 (Section 6.2) it can be written as

$$u_3 = 0.25 r \beta \overline{E} MV. \tag{13.6}$$

The diffraction efficiency ϵ is given by the ratio of the power in the holographic image u_3^2 to that of the illumination r^2:

$$\epsilon = 0.25^2 (\beta \overline{E} MV)^2. \tag{13.7}$$

Since $\beta = dt/dE$ the result is for $\beta E = dt/d \ln E = \log_{10} e^{(dt/d \ln E)} = 0.43\Gamma$. Thus the efficiency is

$$\epsilon = \frac{1}{16} (0.43 \Gamma MV)^2. \tag{13.8}$$

The parameter Γ describes the ascending slope in the t–log E curve (Fig. 13.1). The equation means that the diffraction efficiency is proportional to the sensitivity of the layer given by β or Γ and also to the visibility of the grating V

and to the visibility transfer function M. Here it has to be pointed out that the equation was derived for thin amplitude gratings whose maximum diffraction efficiency is $\epsilon = 1/16 = 6.25\%$ (Chapter 6). This theoretical value cannot be reached because of the transmission being $t < 1$.

13.3
Noise of Emulsion Layers

Due to the granularity of photosensitive layers diffraction and scattering of light occur, which in holography leads to so-called noise [3]. These disturbances are especially important for holograms with low diffraction efficiency and for multiple recordings of holograms on top of each other. In terms of Fourier analysis the irregular structures in the film can be thought of as a sum of diffraction gratings and they therefore represent a spectrum of different spatial frequencies inside the layer. One therefore talks of the "noise spectrum" of the layer.

To measure the noise spectrum a setup according to Fig. 13.3 can be used. Before the investigation the holographic layer is irradiated with a homogeneous light energy $\overline{E}$ such that after developing a certain darkening – or for layers of phase holograms a certain phase shift – occurs. The light energy $\overline{E}$ is roughly the value that is usually used for a holographic recording. The so treated layer is investigated by illuminating it perpendicularly with a homogeneous light wave. If the layer is perfectly homogeneous, i.e., without any granularity, a central (diffraction limited) light spot would appear in the focal plane of the lens in Fig. 13.3. However, if the light is diffracted by the structures of the layer, light also appears to the side of the focal point. The

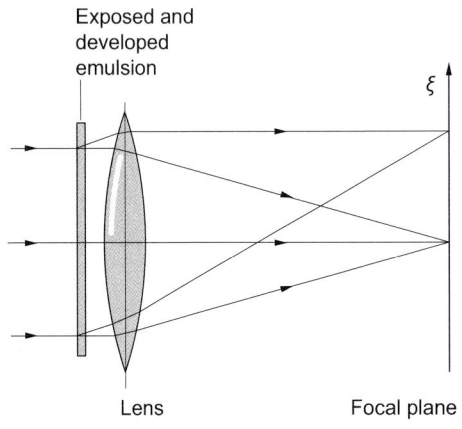

Fig. 13.3 Setup for the measurement of noise of holographic layers. The noise spectrum is measured in the focal plane of the lens.

intensity distribution of the light in the focal plane is therefore a measure for the granularity or the noise of the holographic material.

13.3.1
Fourier Analysis

The spatial frequency spectrum of the photo layer according to Fig. 13.3 appears in the focal plane of the lens. This can be interpreted as follows: if a grating with the spatial frequency $\sigma = 1/d_g$ (d_g = grating constant) exists on the photo layer, parallel light is diffracted at an angle of $\sin\delta = \lambda/d_g = \lambda\sigma$. A spot of the image appears at the position $\xi = f\tan\delta$ in the focal plane of the lens. Here f is the focal distance. For small angles one can assume that $\tan\delta = \sin\delta = \delta$. Thus $\xi = f\lambda\sigma$, i.e., the position of the image point in the focal plane despite a constant factor $f\lambda$ is equal to the spatial frequency. Hence the intensity distribution in the focal plane represents the spectrum of spatial frequencies of the layer.

13.3.2
Measurement Procedure

The noise, i.e., the spectrum of spatial frequencies of a holographic layer, is measured according to Fig. 13.3. The light intensity in the focal plane of the lens is measured using a photodetector. The coordinates in this plane are denoted by ξ and η. The area of the detector measures $\Delta\xi$ and $\Delta\eta$; the according spatial frequency intervals are $\Delta\sigma_\xi = \Delta\xi/\lambda f$ and $\Delta\sigma_\eta = \Delta\eta/\lambda f$.

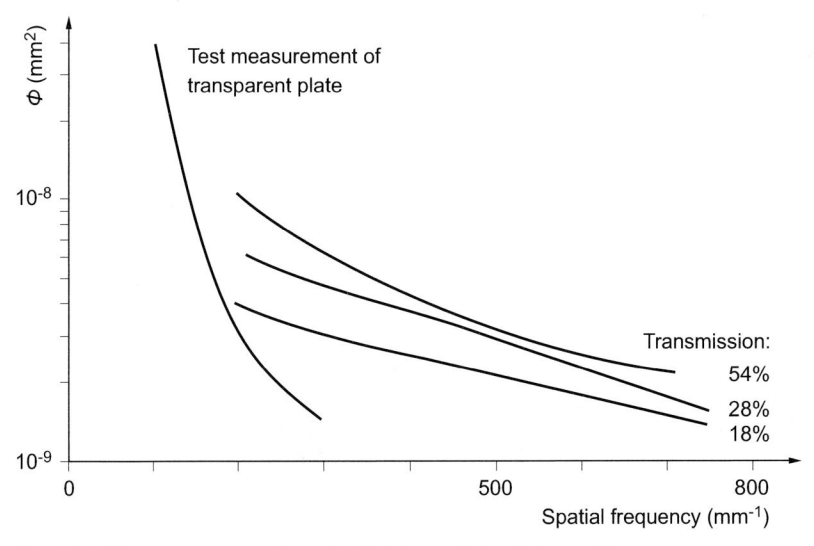

Fig. 13.4 Power spectrum of a typical high resolution holographic emulsion.

The power spectrum $\Phi(\xi,\eta)$ is defined as the power of the noise P_R divided by the spatial frequency intervals $\Delta\xi$ and $\Delta\eta$; in addition a division by the incident light power is P_e is done [3]:

$$\Phi(\xi,\eta) = \frac{P_R}{P_e \Delta\xi \Delta\eta / (\lambda f)^2}. \tag{13.9}$$

The area of the detector $\Delta\xi \cdot \Delta\eta$ is chosen large enough so that speckles are averaged out. The value of Φ is calculated dependent of the radius $\rho = \sqrt{\xi^2 + \eta^2}$ and is drawn as a function of $\sigma_\rho = \rho/\lambda f$, i.e., the spatial frequency, in the radial direction. The parameter Φ is also called the "Wiener spectrum" or "power spectrum." The power spectrum Φ of a holographic layer is shown in Fig. 13.4.

13.4
Nonlinear Effects

For linearly recorded holograms four light waves appear during reconstruction (u_1 to u_4, see Section 2.3): u_1 represents the illumination wave weakened by the hologram which is surrounded by a halo (u_2); u_3 describes the normal holographic image, and u_4 the conjugated one.

In Section 13.1 it was introduced that linear recording is approximately possible for amplitude holograms. Phase holograms are nonlinear in principle since the phase appears in the exponent of an e function.

Nonlinearity causes that for the recording of a $\sin x$ or $\cos x$ grating higher harmonics appear, i.e., spatial frequencies with twice or triple, etc., the value. The amplitudes of these harmonics decrease with rising frequency such that mainly the first harmonic has an influence on the holographic image.

13.4.1
Influence of Harmonics

A doubling of the spatial frequency of a grating causes the diffraction to appear at nearly double the angle. From this it can be derived how the individual waves u_1 to u_4 are influenced by the harmonics. The weakened illumination wave u_1 does not undergo any change; however in addition to the halo u_2 another halo appears at twice the angle. Also in addition to the normal and the conjugated image u_3 and u_4 images appear at twice the angle whose curvature is also doubled. Also new halos appear around the images u_3 and u_4. They have a similar effect as the noise of the emulsion and are caused by the spatial frequencies of the object which are also doubled (according to Fig. 2.6). Details on the influence of nonlinear recording can be found in Refs. [3,8].

13.4.2
Thick Holograms

The appearance of harmonics due to nonlinearities decreases the quality of the recording, especially for thin holograms. For thick holograms the influence is not so serious since diffractions of higher orders are prevented by the Bragg condition of the grating planes.

Problems

Problem 13.1 Assumed a developing process of holograms produces a pure amplitude hologram with a transmission of 50% and a pure phase hologram with the same transmittance. Calculate (Eq. (13.1a)) $\Delta\alpha$ for $\Delta\phi = 0$ and $\Delta\phi$ for $\Delta\alpha = 0$. The emulsion thickness is 8 µm; the glass plate is complete transparent. Calculate the refractive index swing, Δn, too (Eq. (13.1b)).

Problem 13.2 In Fig. 13.2 two examples of storing a grating within a holographic emulsion are given. What differences are observed inspecting a white light point source through both gratings? Explain the differences using Fourier transform theory.

Problem 13.3 Show that in Fig. 13.2a the visibility during exposure must be 1 and in Fig. 13.2b smaller than 1.

Problem 13.4 Figure 13.4 shows the Wiener spectra or power spectra of noise measurements of a transparent plate and some emulsions. What are the optical depths for the three given transmissions? Calculate at $\Phi = 4 \times 10^{-9}$ mm² for the transparent plate and the emulsion with 28% transmission the spatial frequencies from the figure. Derive the grating constant, which can be explained as the mean distance of particles producing the measured noise.

14
Recording Media for Holograms

Suitable media for a holographic recording should exhibit a high sensitivity for the laser wavelengths used, a high resolution, a linear recording behavior, low noise, the possibility of erasing the recording, and reuse it as well as a low price. Depending on the application area different recording media can be used which are listed in Table 14.1 [8].

Tab. 14.1 Overview of holographic recording media [8].

Material	Process	Type of hologram	Spectrum (nm)	Exposure ($\mu J/cm^2$)	Resolution (lines/mm)	Diffr. eff.
Not erasable						
Silver halide	Developing	Amplitude	400–700	1–100	3000–7000	0.05
Silver halide	Bleaching	Phase	400–700	1–100	3000–7000	≥ 0.5
Dichrom.	Developing	Phase	350–580	10^4	>10,000	0.90
Photoresist	Developing	Phase	UV–500	10^4	3000	0.30
Photopol.	Exposure	Phase	UV–650	10^3–10^6	200–1500	0.9
Erasable						
Photochr.	None	Amplitude	300–700	10^4–10^5	>5000	0.02
Thermopl.	Ch., Heat	Phase	400–650	10	500–1200	0.3
Photorefr.	None	Phase	350–500	10^4–10^5	>1500	0.2
"+Photol.*	None	Phase	350–500	10^2	>10,000	0.25

* see Table 14.6

14.1
Silver Halide Emulsions

For decades silver halide emulsions have been used as a photographic film material; in holography these are also the most widely used recording media. They exhibit a high sensitivity and can be sensitized for the desired laser wavelengths by deposition of dyes. These emulsions are used in laboratories that produce artistic or graphic works. In the past few years many manufacturers such as Agfa, Ilford, and Kodak gave up production of holographic emulsions. There are still some Agfa products around, but Agfa does not sell the former 8E56/8E75 and 10E56/10E75. Compared to photography

Holography: A Practical Approach. Gerhard K. Ackermann and Jürgen Eichler
Copyright © 2007 WILEY-VCH Verlag GmbH & Co. KGaA, Weinheim
ISBN: 978-3-527-40663-0

the silver halide crystals in holographic emulsions are much smaller (30 to 90 nm diameter) such that the resolution increases from about 100 to more than 5000 lines/mm.

14.1.1
Working Principle

Silver halide layers for holography usually consist of a 5 to 7 μm thick layer of neutral gelatin which is applied to a glass or film substrate. In this a suspension of silver halide crystals, mostly AgBr, is deposited. By adding heavy metal ions and a weak reaction with sulfide ions the layer is sensitized, i.e., it is made photosensitive. By using special color additives a sensitization for different wavelength areas can be accomplished (see Table 14.2 and Fig. 14.1). During exposure of the emulsion silver is created according to the following equation:

$$AgBr + hf = Ag + Br.$$

During exposure only single Ag seeds are produced inside the AgBr grains which act as a catalytic center. A so-called latent image is created. During the developing process the exposed grain is then completely reduced to silver producing an "amplification" of a factor of 10^6. That way these emulsions are much more sensitive than others (Tab. 14.1). The reduced silver absorbs light and the emulsion appears black. Due to this amplitude holograms are created; these can be transformed into phase holograms by bleaching.

14.1.2
Resolution

For volume holograms the grating constant d_g inside the holographic layer can be calculated from the Bragg condition:

$$d_g = \frac{\lambda_n}{2 \sin\left(\frac{\delta}{2}\right)}, \tag{14.1}$$

where δ denotes the angle between the object and reference wave inside the layer whilst $\lambda_n = \lambda/n$ denotes the wavelength in the emulsion with refraction index n. For example, this yields for typical transmission holograms with $\delta = 60°$, $\lambda = 633$ nm, and $n = 1.64$ a spatial frequency of $\sigma = 1/\lambda_n = 2590$ lines/mm. For white light reflection holograms with $\delta = 180°$ the spatial frequency with $\sigma = 5180$ lines/mm is twice as high. Common photographic layers cannot be used for holography due to their low resolution ($\sigma = 40$ to 600 lines/mm). The carrier frequency for special holographic layers is around 5000 lines/mm (see Table 14.1). Some of the new emulsions shown in Table 14.2 are still in the process of development [28–30]. Therefore within few years the properties of available holographic materials may change. Today the Genet material has the highest resolution. As a consequence the exposure is comparatively high. Very much used in holographic laboratories is Slavich PFG-01 and Fuji F-HL. Hypersensitization using TEA is always necessary.

Tab. 14.2 Properties of holographic silver halide emulsions.

Type	Company	Spectral range/(nm)	Color	Exposure ($\mu J/cm^2$)	Grain size (nm)	Resolution (lines/mm)
Ultimate 15	Gentet	510–660 as specified	Green–Red	~100	15	>7000
Ultimate 08		440–710 as specified	Green–Red	~400	5–8	>10000
VRM-M	Slavich	480–530	Green	20–40 pulse 60–80 cw	40	3000
PFG-01		600–660	Red	80	40	3000
PFG-03M		600–680	Red	1500–2000	10	5000
PFG-03C		450–650	Pan	2000 (blue) 3000 (green, red)	10	5000
BB450	Color holographic	410–470	Blue	150–300	20–25	
BB520		480–540	Green	150–300	20–25	
BB640		580–650	Red	150–300	20–25	
BB700		694	Red, pulse		50–60	
F-HL	Fuji	400–700	Pan	100–200	30–40	>3000
DESA	TFH Berlin	630	Red	80–120	15	5000

14.1.3
Spectral Resolution

The sensitivity of AgBr layers depends on the size of the light sensitive grains which governs the degree of resolution. This means that high resolution films exhibit a low sensitivity (Table 14.2); it varies strongly with the laser wavelength. Depending on the type of laser used it is advisable to to use different emulsions. For the red (He–Ne laser and ruby laser) and the blue (argon and frequency-doubled neodymium laser) region special film material is available. The spectral sensitivity is shown in Fig. 14.1, which illustrates the required energy density for typical holograms using different wavelengths.

Note that the red sensitive films have a gap in the green region whilst the blue/green sensitive are insensitive for red light. This makes it possible to work in the darkroom using the complimentary color as illumination.

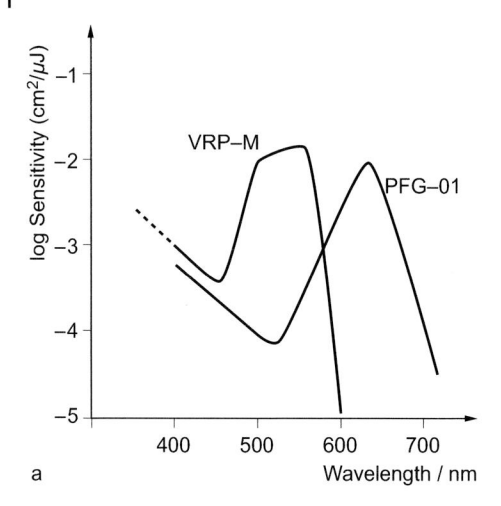

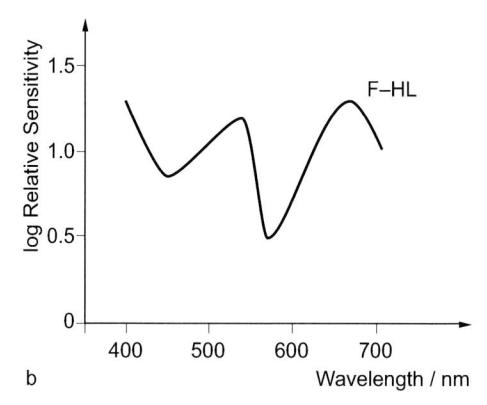

Fig. 14.1 Spectral sensitivity of different holographic silver bromide films. (a) the energy which creates an optical density of $D = 2$ (using a special developing process), films: VRP-M and PFG-01 of Slavich; (b) relative sensitivity for F-HL of Fuji.

14.1.4
H&D Curves

In Fig. 14.2 the optical density of Slavich emulsions VRP-M and PFG-01 is given as a function of exposure. A density of 2 or a transparency of 1% is necessary for white light reflection holograms. For PFG-01 an exposure of about 110 µJ/cm² is necessary to reach $D = 2$. The exact amount depends on the developing process.

14.1.5
Diffraction Efficiency

Typical diffraction efficiency curves are given in Fig. 14.3. VRP-M and PFG-01 of Slavich are shown again. The highest diffraction efficiency is reached at 110 µJ/cm² for PFG-01 and about 80 µJ/cm² for VRP-M. These values correspond with an optical density of 2 (see Fig. 14.2).

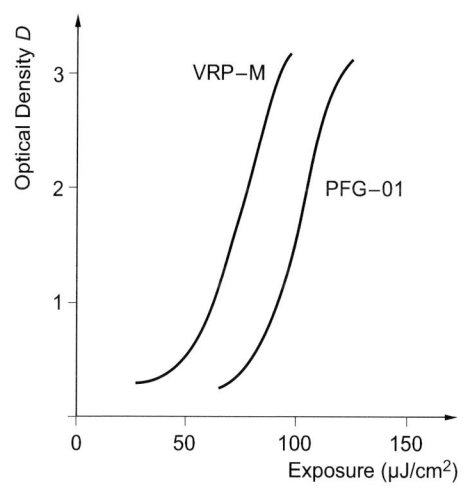

Fig. 14.2 Optical density versus exposure. An optical density of $D \approx 2$ is necessary for white light reflection holograms.

14.1.6
Scattered Light

In Section 13.3 (noise) it was said that the granularity of the AgBr crystals produces scattered light. Fine-grained films exhibit lower scattering than coarse-grained; the scattering occurs mainly for small angles, i.e., for low spatial frequencies (Fig. 13.4). In general the scattering decreases with increasing wavelength; blue light is scattered stronger than red light. During the bleaching process of amplitude holograms to transform them into phase holograms the diffraction efficiency increases but so does the noise.

14.2
Exposure, Developing, and Bleaching

In the following section working with silver halide films will be explained, i.e., exposure process, developing, and bleaching. The films are exposed and developed and in doing so an amplitude hologram is created in the first step.

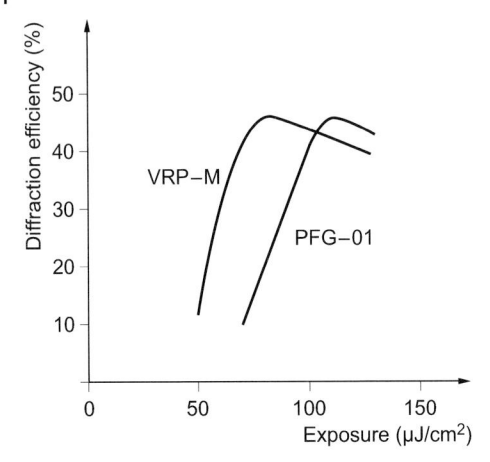

Fig. 14.3 Diffraction efficiency versus exposure, given for VRP-M and PFG-01 of Slavich, respectively. Maximum diffraction efficiency is reached for 80 µJ/cm² and about 110 µJ/cm², respectively.

Amplitude gratings have only a low diffraction efficiency (7% for volume reflection holograms), so that in a second step a transformation into a phase hologram is necessary. The diffraction efficiency in this case can reach 100%; very brilliant images can be created. In the following different processes during the transformation of amplitude into phase holograms will be explained.

14.2.1
Exposure

The intensity ratio of reference and object waves is inconsistent and varies from $I_r/I_o = 1:1$ to 10:1 [29]. In single-beam holography this ratio depends only on the reflection efficiency of the object and cannot be adjusted. When selecting the intensity ratio one has to distinguish between thin and volume holograms. For thin holograms one has to pay attention that the variation of the intensity does not become too large so that the linear region of the H&D curve is not left. An intensity ratio of $I_r/I_o = 4$ to 10 is recommendable. According to Fig. 13.2b the linear region of the H&D curve can be reached by a constant light background. The visibility V of the interference fringes for $I_r/I_o = 10$ can be calculated by

$$V = 2\frac{\sqrt{\frac{I_r}{I_o}}}{1 + \frac{I_r}{I_o}} \qquad (14.2)$$

to $V = 0.57$. According to Fig. 13.2b this value is a good result since the light grating deviates by ±57% from the mean value. Deviations in the intensity of the object wave which can be very large due to the geometry and the reflection ability of the object in this case do not influence the total intensity that much. Getting into the nonlinear region of the H&D curve would cause increased noise, halo effects, and ghost images due to higher order diffractions.

Note that according to Eq. (14.2) the visibility does not depend that much on I_r/I_o because during interference amplitudes and not intensities superpose. By reducing the visibility from its maximum value of 1 down to 0.57 the diffraction efficiency of thin amplitude gratings changes according to Eq. (13.8) by a factor of $0.57^2 = 0.32$. Similar holds for thin phase gratings because the resulting phase difference caused by the bleaching is proportional to the darkening. It has to be noted though that in general phase gratings exhibit a nonlinear behavior (Section 13.4).

For volume holograms different relations are valid for the intensities of reference and object waves because due to the Bragg condition higher diffraction orders do not occur. Therefore in this case much lower intensity ratios, such as 3:1 or 1:1, are chosen. Due to the nonlinearity of the photographic layer halo effects can occur according to the theory of coupled waves (Section 6.3).

14.2.2
Phase Holograms

For the description of thick phase holograms the modulation parameter Φ is introduced (Section 6.3):

$$\Phi = \frac{\pi n_1 d}{\lambda \cos \delta}. \qquad (14.3)$$

Here d denotes the thickness of the emulsion, δ the angle of incidence of the reconstruction wave with respect to the grating plane and n_1 is the refraction index difference between exposed and unexposed areas of the phase grating.

The theory shows that for transmission holograms the modulation parameter should be $\Phi \approx \pi/2$ to obtain the highest diffraction efficiency. For values above $\pi/2$ it again decreases. For reflection holograms the diffraction efficiency almost reaches 1 if $\Phi \approx \pi$ and if the illumination obeys the Bragg condition. (Above this value the diffraction efficiency also increases outside the Bragg angle.) This results in the fact that the parameter n_1 needs to be larger for reflection holograms than for transmission holograms. These theoretical findings about phase holograms only contain very indirect hints for the practice. The optimal parameters for exposure, developing, and bleaching can only be determined experimentally.

14.2.3
Optical Density

In holography exposure and developing times are chosen such that the holograms obtain a maximum diffraction efficiency. For amplitude holograms the corresponding optical density D after the developing can be calculated. For modulations around the center of the H&D curve the resulting mean value is $t = 0.5$. Since no full modulation is possible the real optimal value is $t = 0.45$. For transmission holograms follows: $D = \log 1/t^2 = 0.7$.

For phase holograms the desired optical density after developing can only be determined experimentally. However, the optimal phase difference in the holographic grating can be calculated.

Due to the required large modulation parameter and the necessary large refraction index difference D will be chosen as large as possible for reflection holograms. For densities above $D = 2$ the noise increases. Therefore, this value should not be exceeded. An exact determination of the optical density can only be done in the experiment.

14.2.4
Phase Holograms by Bleaching

To increase the diffraction efficiency amplitude holograms are converted into phase holograms after developing. Two methods for this exist:

(1) The phase hologram is created by a variation in thickness of the emulsion. This technique is not very common for silver halide emulsions since the maximum reachable diffraction efficiency is only 33.9%. It is used however for thermoplastic films in holographic instant cameras and for photoresist layers, e.g., for the production of embossed holograms.

(2) By bleaching phase differences are created inside the layer which contain information. Here three different methods are used which are based on different chemical reactions (Fig. 14.4).

In the *conventional* bleaching process the hologram layer is developed and fixed. "Fixing" means that the unexposed silver halide is removed from the gelatin. This way an amplitude hologram is created containing areas with and without silver (Fig. 14.4a). During the bleaching the silver is transformed back into transparent silver halide. This causes the index of refraction in these areas to be larger than in the gelatin and a phase hologram is created. The structure of the transformed silver halide is almost insensitive to the photochemical reaction described in Eq. (14.1) since it is not sensitized.

The *reversal bleaching* or solving bleach omits the fixing process (Fig. 14.4b). After developing fringes with Ag and AgBr are created. The bleaching bath removes the exposed Ag from the emulsion layer. The resulting phase holo-

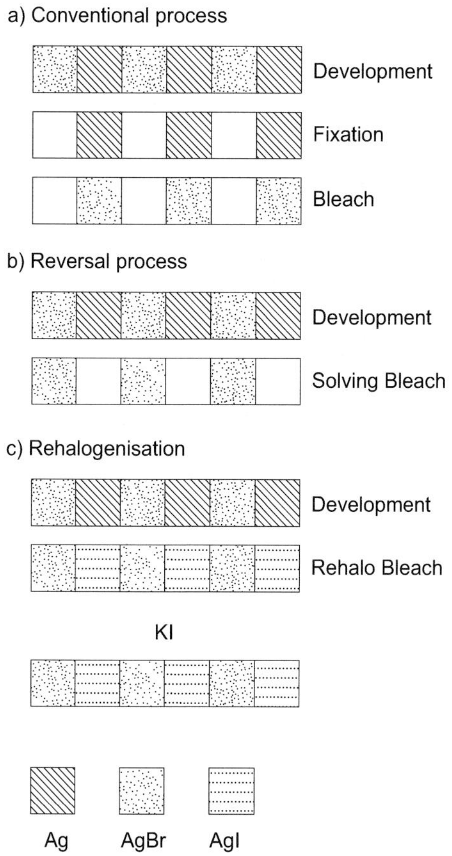

Fig. 14.4 Schematic overview of the different bleaching processes: (a) cconventional process, (b) reversal process, and (c) rehalogenisation.

gram is similar to that in Fig. 14.4a but exposed and unexposed areas are exchanged. For the holographic reconstruction however this makes no difference. In both bleaching methods material is removed from the emulsion layer so that it shrinks.

The shrinkage is avoided almost completely by the *rehalogenizing bleaching* (Fig. 14.4). After the developing the exposed silver is transformed back into AgBr by the bleaching bath (Fig. 14.4c). One would expect that now all the information is destroyed because now the complete layer contains AgBr. However, during the process the rehalogenized silver diffuses into the unexposed areas which already contain AgBr seeds. This way the concentration of AgBr varies and a phase grating appears. As shown in Fig. 14.4c the exposed Ag can also be turned into AgI (instead of AgBr). To understand the stability of the layers against light after the bleaching it has to be noted that unexposed AgBr is desensitized by watering.

14.2.5
Shrinkage of the Emulsions

By removing substances during conventional and reversal bleaching a shrinkage of the photo layer of about 15% occurs. If the grating lines are across the layer, as in Fig. 14.4, mainly a surface structure occurs. Such structures are also caused by drying (Fig. 14.5).

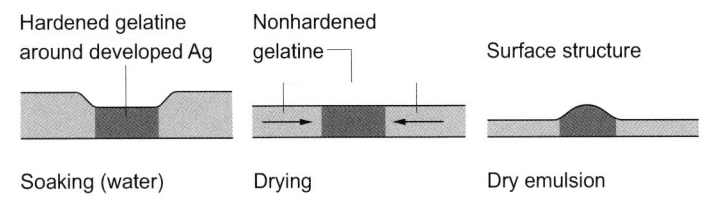

Hardened gelatine around developed Ag | Nonhardened gelatine | Surface structure

Soaking (water) | Drying | Dry emulsion

Fig. 14.5 Formation of surface structures during drying of the gelatin layer.

During the developing process the layer thickness increases by a factor of 5 due to water soaking. During this the areas around the developed silver grains soak up less water. Due to mechanical tensions the gelatin is pushed into the areas of the Ag grains during the drying such that the thickness remains higher there. For transmission holograms this relief structures only have a secondary effect since there spatial frequency is very high. A noise spectrum at small diffraction angles is created.

For white light reflection holograms where the grating planes are parallel to the photographic layer the shrinkage has a different result; It causes a decrease of the grating constant. Therefore, the Bragg reflection occurs for these planes at a smaller wavelength λ'. λ' can be calculated from the recording wavelength λ and the grating constants d_g and d'_g before and after the shrinkage:

$$\lambda' = \lambda \frac{d_g}{d'_g}. \tag{14.4}$$

Because of the shrinkage images recorded with red lasers often appear green during reconstruction. With a rehalogenizing bleaching bath this effect can be almost avoided. Consequently the color of white light holograms can be influenced by the choice of the bleaching bath.

14.2.6
Pseudocolors, Preswelling

The shrinkage can be used for the creation of special color effects for reflection holograms. During the so-called preswelling the emulsion is soaked in TEA (triethanolamin) aqueous solution of different concentrations. The swelling before the exposure causes the grating constant of the dried hologram to be smaller. Therefore, white light holograms change their color into the blue region of the spectrum.

The following recipe describes the creation of defined color shifts. The film is soaked in triethanolamin and dried afterward where only the water escapes from the gelatin layer. When using an EDTA bleaching bath (Table 14.4a) the colors of red (He–Ne) exposed reflection holograms is switched to orange up to violet by setting the concentration of the Triethanolamin in water between 1.3 and 17%.

Also for finished reflection holograms the layer thickness and the grating constant can be influenced. Breathing on the hologram causes a color shift into the red whilst drying with a blow dryer causes a blue shift. A different possibility is to soak the finished hologram in aqueous Triethanolamin and to dry it afterward. Since the triethanolamin remains in the layer and increases it the result is a red shift. This method is called "postswelling."

14.2.7
Index Matching

When putting the holographic film between two glass plates usually layers of air stay between the film and glass. The difference of the indices of refraction at the interfaces causes a reflection of 4% each. The reflected waves form interference structures which are visible on the hologram. To avoid this effect the film is embedded in a liquid between the glass plates. By using this method called "index matching" the refraction indices are equalized and the undesired interferences almost disappear. The applied liquids should have the index of refraction of the glass or the carrier film, and should not be to volatile and odorless. Possible liquids are turpentine replacement, thinner, alcohol, or glycerine. Alcohol evaporates a little bit fast while glycerin is a bit viscous and the edges of the film or the glass plate need to be clamped together [29]. A better procedure to overcome unwanted interferences is a vacuum plate holder. Then stabilizing glass plates can be omitted.

14.2.8
Developer

Developers for holographic emulsions contain a reducing agent, e.g., hydrochinone, methol, pyrogallol, or catechol with additives such as sodium carbonate, sodium sulfite, and chemicals for the preservation of the unstable solution. The alkali is needed to bind the bromine (e.g., NaBr). A very often used developer is CWC2. The recipe is given in Table 14.3b. This developer should be used to process Slavich emulsions.

Tab. 14.3 Recipes for developers.

(a)	Metol	2 g	
	Ascorbic acid (vitamin C)	20 g	1 liter distilled water
	Sodium carbonate	50 g	
(b)	Catechol	10 g	
	Ascorbic acid (vitamin C)	5 g	1 liter distilled water
	Sodium sulfite	5 g	
	Uric acid	50 g	
	Sodium carbonate	30 g	
(c)	Part 1: pyrogallol	5 g	0.5 liter distilled water
	Part 2: soda (Na$_2$CO$_3$)	60 g	0.5 liter distilled water
	Reacts instantly after putting together of parts 1 + 2		
	(brown coloring). Usage once for multiple recordings		

During developing the temperature of the bath should be kept stable to 0.5°C. Rising the temperature to, e.g., 35°C will speed up the developing process but shorten the lifetime of the developer. Bad results after the developing are often caused by developers that are too old or by too long developing times which increase the noise.

The useful life of the developers can be significantly increased if the bath is covered with a glass plate and protected by an inert gas. This way the usability of catechol developers can be extended from 1 day to several weeks. Beneath the mentioned recipes several developers are commercially available.

14.2.9
Bleaching

Amplitude holograms (after developing) show very small diffraction efficiency. Therefore, the holograms should always be transformed to phase holograms. Bleaching the amplitude holograms does this. As mentioned above there exists two basic processes: solving and rehalogenating bleaches.

Recipes are given in Table 14.4. The most commonly used rehalogenating bleach is PBU, which is useful for Slavich as well as for emulsions of color holographics (BB plates). If solving bleach is used, then the information will eventually vanish. This happens after a developing process producing colloidal silver instead of metallic silver. In this case the plates look red or brown after developing.

Rehalo bleach should be taken for white light reflection holograms in order to prevent wavelength shift due to emulsion shrinkage.

For transmission holograms solving bleach is used very often. Potassium dichromate (or ammonium dichromate) is the main ingredient as shown in Table 14.4. The emulsion shrinkage does not reduce the quality of transmission holograms because the interference fringes within the emulsion layer are perpendicular to the emulsion surface.

The plates should always be left within the bleaching solution until they turned completely transparent.

In phase holograms the unexposed AgBr slowly darkens with time. Therefore, the layer needs to be desensitized by soaking in water (15 min). Hariharan suggests to exchange the KBr (4 g) according to Tab 14.4 by the less light sensitive KJ (2 g). The bath may be used only once. Alternatively, a little bit of KJ can be added to the last water bath.

In Table 14.5 the most important steps for developing, bleaching, and desensitizing are summarized.

It is not necessary to wait until the hologram has dried completely in order to get an impression of successful exposure. After developing and/or after bleaching the still wet holographic plate should be illuminated with white light. Also if it is a reflection hologram, some kind of rainbow can be seen in transmission – if everything runs well during exposure. This rainbow is an indication for a stored hologram which will be seen once the plate has dried completely.

Tab. 14.4 Bleaching baths

(a)	Rehalogenating PBU bleach		
	Copper (II) bromide	0.5 g	
	Potassium, peroxidysulfate	5 g	
	Ascorbic acid	25 g	
	Potassium bromide	10 g	
	Distilled water	500 ml	
	After completing the solution add hydrochinone	0.5 g	
or			
	Potassium dichromate	7 g	
	Potassium bromide	4 g	1 liter distilled water
	Sulfuric acid	2 ml	
(b)	Solving		
	Potassium dichromate	5 g	1 liter distilled water
	Sulfuric acid	5 ml	

Exposure, developing, and bleaching of holograms require a lot of experience which can only be acquired by experimenting. The overview of this section should help to understand the most important steps for the production of holograms.

Tab. 14.5 Summary of steps for developing and bleaching of phase holograms.

1. Developing	Optical density depending on the type of hologram, e.g., $D = 2$ for white light reflection holograms
2. Soaking	2 min
3. Bleaching	Wait for complete transparency. Light might be switched on. Protect bath from light: this prevents early darkening which has no effect on the diffraction efficiency
4. Soaking	2 min under running water, 13 min in stagnant water. The emulsion is desensitized. Cover bath, otherwise emulsion will turn gray
5. Soaking	1 min with a few drops of Fotoflo. A few grains of KI may be added
6. Drying	Only in dry state the hologram will appear

14.3
Dichromate Gelatin

Dichromate gelatin has excellent properties as a recording medium for volume holograms: Large differences of the refraction index of the recorded gratings, high resolution as well as low absorption and scattering. The diffraction efficiency is close to the theoretical limit of 100% though its sensitivity is with 10 mJ/cm^2 nearly 1000 times higher than that of holographic AgBr films.

14.3.1
Working Principle

It is known for about 150 years already that blue or ultraviolet light can harden gelatin layers in which a small amount of dichromate is deposited and make them water resistant. By photochemical reactions chromium ions Cr^{3+} are created in the dichromate, e.g., ammonium dichromate $(NH_4)_2Cr_2O_7$, which lead to localized bonds of carbon–oxygen groups of neighboring gelatin chains. Sufficiently exposed areas inside the layer become hard and water resistant whilst the unexposed material can be washed away with warm water. These properties lead to its application as a photoresist material which forms surface structures.

14.3.2
Preparation of Dichromate Gelatin (DCG) Holographic Plates

Slavich company produces DCG for holographic use (PFG-04). Because of limited shelf lifetime (1 month for PFG-04) one can also produce those plates in the laboratory. The emulsion consists of dichromate (ammonium dichromate or potassium dichromate), gelatin and water. The reader may inspect special publications for preparation. The main steps are given here.

First, the gelatin layer is produced on a glass plate using a spin table. A solution of 7% gelatin is taken and the layer dried and hardened. These steps may be repeated until the total thickness is between 10 µm and 20 µm.

In order to shorten the production normal holographic plates can be used. The silver halide is removed by a fixing bath.

For sensitization the plate is soaked for 5 min in a solution of 5% $(NH_4)_2Cr_2O_7$ in water.

The plates are dried and are ready for use.

14.3.3
Properties of DCG Holographic Plates

DCG is sensitive in the green-blue region of the spectrum. Slavich company plates PFG-04 have a maximum sensitivity at $\lambda = 415$ nm. The material like all DCG is of low light sensitivity. Using an Ar-ion laser the exposure is 80 mJ/cm^2 at $\lambda = 475$ nm, 100 mJ/cm^2 at $\lambda = 488$ nm, and 250 mJ/cm^2 at $\lambda = 514.5$ nm. It is advantageous to use a He–Cd laser, emitting at $\lambda = 442$ nm. The sensitivity of DCG is about a factor of 10 higher at this wavelength. The emulsion layers are completely insensitive to the light of the He–Ne laser; however by sensitizing with dyes, e.g., methylene blue or green, the sensitivity can be extended into the red. But the sensitivity is a factor of 50 to 100 lower than at 488 nm.

The behavior of dichromate layers is highly dependent on the hardening process; soft layers are more sensitive to light. However, for poor hardening large pores can occur causing the layer to become milky, which leads to noise. The resolution is about 10,000 lines/mm; the achievable differences in index of refraction are 0.08 quite large.

14.3.4
Exposure and Developing

During the holographic recording the film is exposed with 25 to 250 mJ/cm^2. After exposure the hologram has to be hardened thermally (100°C) and cooled down to room temperature. In order to see the hologram the DCG has to follow some bathing steps in water and alcohol. Within the water the DCG in unexposed areas is washed out. This process is followed by a drying bath in isopropyl alcohol. Slavich company always forward the details.

The holograms have to be sealed to prevent dissolving due to air humidity. Very often a protective glass is used or epoxy glue; DCG holograms are therefore always thicker than silver halide holograms. But they are of low noise, high diffraction efficiency holograms with a very deep and detailed structure.

14.4
Photothermoplastic Films

In the so-called holographic instant cameras photoconducting layers are used together with thermoplastic layers. The developing after exposure is done electronically without removing the film from its original position. This way it is possible to perform real-time holography for interferometric applications.

14.4.1
Structure of the Layers

According to Fig. 14.6a, a thin conducting layer is applied to a substrate of glass or film which is connected to a negative voltage or ground during the recording and developing. Above that is a layer of photoconducting material about 1 µm thick which acts as an insulator in darkness. As a last layer an isolating thermoplastic material about 0.7 µm is applied, which acts as the recording medium and becomes liquid at a temperature of about 70°C. This topmost layer is sprayed with positive ions using a corona discharge during the recording and developing: for the charging process a setup according to Fig. 14.6 is automatically run above the surface. All layers, including the carrier, are transparent. In some films a single layer is used which acts as a photoconductor and a thermoplast at the same time.

In general, the producers of holographic instant cameras also offer thermoplastic layers. Some cameras work with erasable, some with nonerasable layers; the difference is described below. Information about the production in your own laboratory can be found in [3].

Exposure

Before the holographic recording the upper photolayer, i.e., the thermoplastic material, is charged positive. Negative charges, i.e., electrons, cluster in the lower conducting layer on the carrier due to induction (Fig. 14.6b). After that follows the exposure, where for example, a He–Ne laser about 10 to 100 µJ/cm^2 is needed. In the bright areas of the holographic interference pattern the sandwiched photoconductor becomes conducting such that the electrons can travel to the interface with the thermoplastic layer. This is compensating the positive charges (Fig. 14.6b,2). A charge distribution is created that resembles the holographic information. By a second charging procedure additional positive charges can be brought to the exposed areas.

Developing

By heating the thermoplastic, i.e., the recording medium of the hologram, the charge distribution is transformed into a surface relief. A current through the conducting layer of the carrier or an extra heating layer creates temperatures

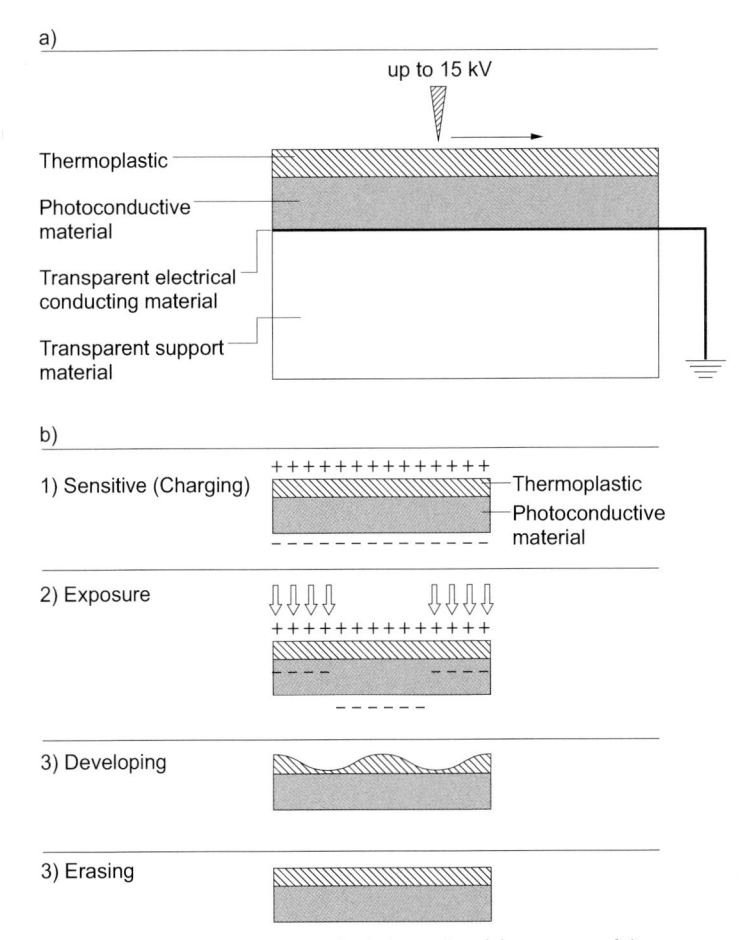

Fig. 14.6 Thermoplastic layers for holography: (a) structure of the layer and illustration of the charging process and (b) function of the layer during sensitizing, exposure, developing, and erasing.

between 60 and 100°C. The material becomes soft and the electrostatic forces push material out of the strongly charged regions. This way the layer becomes thin at the exposed areas and thick at the unexposed (Fig. 14.6b,3). A thin phase hologram is created.

Erasing

In some layers the information can be erased 100 times by reheating the material in the discharged state. The soft material is then flattened again by the surface tension (Fig. 14.4b,4). For erasable layers glass plates are used as a carrier, whilst for nonerasable transparent films are used.

14.4.2
Optical Properties

Instant cameras for holography usually work with layers having an area around 10 cm² which is sufficient for interferometric purposes. The modulation transfer function depends on the layer thickness and the exposure; it reaches a maximum at a spatial frequency of about 1000 lines/mm at 10 µJ/cm². In Fig. 14.7 the diffraction efficiency of the layer is shown as a function of the spatial frequency which is given by the angle Θ between object and reference waves. The optimum diffraction efficiency of about 10% occurs for transmission holograms at $\Theta \approx 30°$. The decrease of the diffraction efficiency at low spatial frequencies has a positive effect on the disturbances caused by the nonlinear properties of thin phase holograms. A problem with thermoplastic films is the noise which looks like a whitish fog and is therefore called "frost."

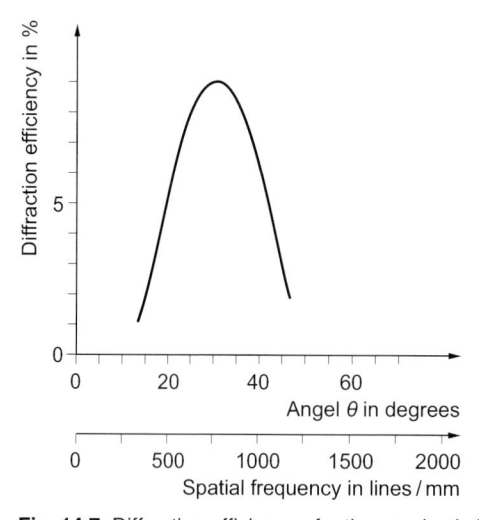

Fig. 14.7 Diffraction efficiency of a thermoplastic holographic film as a function of the spatial frequency and the angle Θ between reference and object waves (He–Ne laser).

14.5
Photoresists

Photoresists are mainly used for the production of holographic optical elements, e.g., gratings or embossed holograms. In the production of integrated semiconductor circuits they are known for a long time. They are applied to a carrier and exposed afterward. Positive photoresists are soluble in developers at their exposed areas, while negative are insoluble. They can be used to create holographic surface structures. Photoresists are relatively insensitive and need over 100 mJ/cm^2 for the exposure. They are used for the production of dies for the duplication of holograms.

Only a few photoresists have a sufficiently high resolution for holography. Mostly used for holographic applications is Shipley "Microposit S1800"-series photoresist [65]. The lacquer is applied to a spinning glass carrier creating layers of 0.5 to 2.6 µm in thickness. These are heated for 15 min at 75°C to completely evaporate the solvent. The layer is especially sensitive in the violet and UV regions such that best matching laser is an argon laser at 488 or 458 nm or even better a He–Cd laser at 442 nm. The nominal sensitivity at 458 nm is at 250 mJ/cm^2 meaning that long exposure times and very stable setups are needed. After exposure the exposed areas are removed, e.g., with developer AZ-303. Within certain limits the etching rate is proportional to the exposure time; a thin phase hologram is created. Often these holograms are further processed, e.g., to produce mirrored gratings, other holographic optical elements or embossed holograms.

14.6
Other Recording Media

14.6.1
Photopolymers

Photopolymers are organic substances that polymerize under the influence of light, i.e., form long molecule chains or other bonds. By the light and a following chemical treatment the index of refraction changes. Thick layers for phase holograms can be produced which have a high diffraction efficiency of 85% and a very good angle selectivity.

Main producer of photopolymer products is DuPont company [31]. The first film was OmniDex 706, followed by HRF and today some experimental films. The highest light sensitivity is around 500 nm (458–528 nm). Therefore, the material is well suited for Ar-ion lasers as well as double frequency Nd–YAG. The necessary exposure is about 40 mJ/cm^2.

The advantage of a polymer holographic material is the ease of handling after exposure and the high diffraction efficiency compared to DCG. In order to fix the image only UV light and heat are necessary. No chemical developing is necessary.

Unfortunately, DuPont is not very much interested in selling this material to single holographers. The company is more interested in high volume users. The experience with polymer is therefore limited.

14.6.2
Photochromic Material

Photochromic materials change their color when exposed to light. Up to now organic photochromics lose their information over time so inorganic recording media, especially doped crystals, are preferred. They have no grain and therefore a high resolution. Due to the large thickness several holograms can be recorded on top of each other. No developing is needed after exposure and the holograms are erasable. However, the applications are limited due to the low diffraction efficiency of about 2%.

A diffraction efficiency that is higher by a factor of 10 can be found in photodichroitic crystals, i.e., doped alkali halides. The holographic recording is done with linear polarized light; only one laser is needed for recording, reading, and erasing the information because the two perpendicular directions of polarization are used.

14.6.3
Photorefractive Crystals

Some electro-optical crystals change their index of refraction (about a factor of 5×10^{-5}) when exposed to light. This change is reversible by using heat or light. Hence it is possible to achieve an erasable three-dimensional holographic recording of information. Due to the Bragg condition several holograms can be recorded on top of each other and be reconstructed. Problems are still with reading the information without erasing the hologram. The most important crystals are shown in Table 14.6; very often lithium niobate ($LiNbO_2$) is used with which a thermal fixing at temperatures between 200 and 300°C is possible. However, the needed power density for the exposure is quite high (0.3 J/cm² at 1% diffraction efficiency); a higher sensitivity is given by BSO and KTN.

Tab. 14.6 Properties of different photorefractive crystals. The energy density corresponds to a diffraction efficiency of 1%. For some photoconducting crystals an external electrical field is required.

Crystal	Wavelength (nm)	Energy density (mJ/cm²)	Electric field (kV/cm)	Remarks
$LiNbO_3$:Fe	488	300		
	351	200	15	
$LiTaO_3$:Fe	351	11	15	
$BaTiO_3$	458	1–10	10	
$Sr_xBa_{1-x}Nb_2O_6$:Ce	488	1.5		Low storage time
$Bi_{12}SiO_{20}$ (BSO)	514	0.3		
$KTa_{1-x}Nb_xO_3$ (KTN)	350	0.05–0.1	10	Low crystal quality

Problems

Problem 14.1 Could one use the emulsion Slavich VRM-M to make a single beam white light reflection hologram, using an Ar laser and $\lambda = 514$ nm? What is the maximum angle between reference and object waves for this emulsion (refractive index $n = 1.5$)?

Problem 14.2 What resolution of holographic material is needed to make a white light reflection hologram with the object wave impinging the holographic plate normal to the surface and the reference wave at an offset angle of 30° with respect to the normal of the plate (other side, of course). Calculate for the He–Ne, wavelength $\lambda = 632.8$ nm and $n = 1.5$.

Problem 14.3 Using a He–Ne laser ($l = 632.8$ nm, $P = 10$ mW) and Slavich PFG03-M to make a white light reflection hologram, calculate the minimum exposure time. The set up is given in Fig. 10.2. The visibility was measured to be 70%. Assume that the divergence of the reference wave is just the size of the holographic plate (10×10 cm^2)

Problem 14.4 After bleaching a white light reflection hologram made with a He–Ne laser ($\lambda = 632.8$ nm) the emulsion shrank by 20%. Calculate the wavelength of the reconstructed image.

Problem 14.5 A white light reflection hologram made with a He–Ne laser (632.8 nm) shows after developing and bleaching a green color. By aspirating the emulsion side of the hologram the color turns to red, warming up the hologram turns the color to green or even to blue. Explain what happened.

Problem 14.6 What is the reason of higher resolution of dichromate gelatine compared with silver halide emulsions?

Problem 14.7 Soaking a white light reflection hologram and a transmission hologram for a while into a triethanolamin bath of some percentage (say 5%) what will generally happen to the reconstructing wavelength of the two holograms after drying?

Part 4   Application of Holography

15
Holographic Interferometry

The holographic interferometry is used in technology to measure small deformations of objects. With the development of lasers and the holography this technique is of growing importance compared to the known methods of photoelasticity and the Moiré technique. The advantage compared to photoelasticity is that small deformations (<1 µm) of diffuse reflecting objects can be measured. This way the holographic interferometry plays an important role in the nondestructive material testing. Static and dynamic holographic methods are used in the automotive industry to measure stress and strain, deformations and vibrations qualitative and quantitative.

In this chapter the most important basic techniques of holographic interferometry are presented and a short insight into the quantitative analysis of three-dimensional interferograms is given. For a more in-depth understanding of the quantitative analysis as well as digital methods the reader is referred to the literature [32].

15.1
Double-Exposure Interferometry

15.1.1
Principle

In a double-exposure interferogram two holographic recordings of the object are stored in one photographic layer: A first one in undisturbed state and a second with a slightly deformed object. During the reconstruction both object waves are "read out" and they interfere with each other. This creates clearly visible interference fringes which cover the whole object. The distance of two bright fringes correspond to a phase shift of 2π or λ, respectively. The density of the fringes shows the spatial distribution of the deformation.

For very large deformations of more than 100 µm then the distance of the fringes becomes so small that an evaluation is almost impossible. The lower limit of recognizable disturbances is fractions of a wavelength, i.e., when using He–Ne lasers well below 0.1 µm.

Holography: A Practical Approach. Gerhard K. Ackermann and Jürgen Eichler
Copyright © 2007 WILEY-VCH Verlag GmbH & Co. KGaA, Weinheim
ISBN: 978-3-527-40663-0

15.1.2
Theory

The object wave is given by Eq. (2.4):

$$\mathbf{o}(x,y) = o(x,y)e^{-i\Phi_0}. \tag{2.4}$$

The slightly deformed object wave is

$$\mathbf{o}'(x,y) = o(x,y)e^{-i\Phi_1}. \tag{2.4'}$$

The very small deformation only affects the phase Φ. The amplitude transmission $t(x,y)$ is given by Eq. (2.22a) with half of the exposure time $(t/2)$ for the two images. The addition of both equations yields

$$t(x,y) = C_1 + C_2[(\mathbf{o}+\mathbf{o}')\mathbf{r}^* + (\mathbf{o}+\mathbf{o}')^*\mathbf{r}] \quad \text{with} \tag{2.22a'}$$

$$C_1 = t_0 + C_2\left(2r^2 + o^2 + o'^2\right) \quad \text{and} \quad C_2 = \frac{\beta\tau}{2}.$$

From Eq. (2.22a') only the part $t_i(x,y)$ of the amplitude transmission is considered which is important for the interferometry:

$$t_i(x,y) = C_2\left(\mathbf{o}+\mathbf{o}'\right)\mathbf{r}^*. \tag{15.1}$$

Reconstructing with the reference wave yields

$$\mathbf{u}_i(x,y) = C_2 r^2\left(\mathbf{o}+\mathbf{o}'\right). \tag{15.2}$$

The observed intensity $I_i = |\mathbf{u}_i(x,y)|^2$ is

$$I_i(x,y) = C\left(\mathbf{o}+\mathbf{o}'\right)\left(\mathbf{o}+\mathbf{o}'\right)^*, \quad \text{with} \tag{15.3}$$

$$C = C_2^2 r^4. \tag{15.4}$$

From Eq. (15.3) results the intensity of the interference pattern:

$$I_i(x,y) = C(o^2 + o'^2) + C(\mathbf{o}'\mathbf{o}^* + \mathbf{o}'^*\mathbf{o}) \tag{15.5}$$

$$= 2Co^2 + Co^2\left(e^{i(\Phi_0-\Phi_1)} + e^{-i(\Phi_0-\Phi_1)}\right) \tag{15.6}$$

$$= C^*\left[1 + \cos(\Phi_0 - \Phi_1)\right]$$

$$I_i(x,y) = 2C^*\cos^2\left[\frac{\Phi_0 - \Phi_1}{2}\right] \quad \text{and} \tag{15.7}$$

$$C^* = 2o^2 C.$$

In the area of reconstructed image the intensity is modulated with $\cos^2(\delta/2)$, $\delta = \Phi_0 - \Phi_1$ according to Eq. (15.7). The brightness of the image becomes 0 if the phase difference of the two object waves is an uneven multiple of π:

$$I_i = 0 \quad \text{for} \quad \delta = (2m+1)\pi \quad m = 0,1,2,\ldots.$$

It becomes maximal if δ is an even multiple of π:

$$I_i = 2C^* \quad \text{for } \delta = 2m\pi \quad m = 0,1,2,\ldots.$$

The image of the object is covered with bright and dark fringes whose distance corresponds to a phase difference of 2π or λ between two adjacent undisturbed or two disturbed object areas, respectively.

15.1.3
Practical Realization

For the realization of a double exposure all setups, which were mentioned in Chapters 9 and 10 may be used. Reflection as well as transmission setups can be applied. For reflection holograms the developing process often causes a shrinkage of the emulsion as well as an according shift in the reconstruction wavelength. However, this effects both recordings in the same way. Disturbing for the double exposure is a movement of the object or the holographic components between the two recordings which causes an undesired interference effect. The evaluation of the interferogram is ambiguous with respect to the direction of the measured deformation of the object.

15.1.4
Sandwich Method

In the past Abramson proposed the sandwich method in order to overcome these problems for not to complicated objects.

The essential part is that not only both recordings of the double exposure are performed after each other as mentioned above but that they are done on different plates at the same place. To evaluate the interference effect both plates are reconstructed as a sandwich.

Since it is not possible to have both plates at the same place during the later reconstruction from the beginning two plates are exposed as a sandwich for each recording, with and without deformation. That way it is made sure that during the reconstruction of two plates from two different recordings the geometry and the path of light are still the same for both recording situations.

The procedure is rather complicated and therefore it is not applied very often. The advantage of the sandwich method is that the direction of the deformation can be determined by tilting the sandwich until the interference fringes vanish [33].

15.2
Real-Time Interferometry

Much larger flexibility for the observation of interference effects is offered by real-time interferometry. The interferometric phenomena can be seen while they are evolving and the effects of changes made to the experimental setup when testing an object can be observed immediately.

15.2.1
Principle

The illuminated object is viewed through a hologram which was made from the undisturbed object before the experiment and that is replaced to its original position during the recording. Consequently, our eye is hit by the reconstructed object wave **o** as well as the scattered object wave **o'**. Figure 15.1 shows the imagined path of the object waves **o** and **o'** to the right of the hologram. The difference is exaggerated here. On the left side only the object wave **o'** exists, i.e., the scattered light from the deformed object.

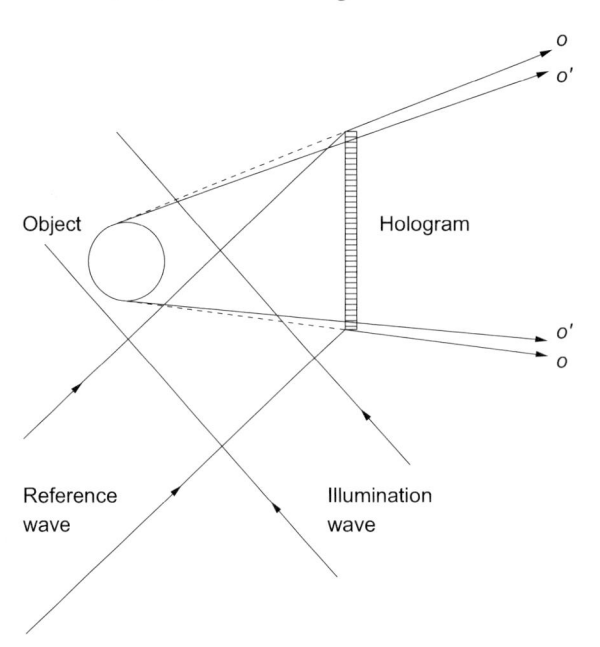

Fig. 15.1 Real-time holography. The undisturbed object wave o is read out from the hologram by the reconstruction wave r. It is then superposed by the object wave o' which is scattered from the deformed object directly.

15.2.2
Phase Difference Between o and o'

The hologram plate is illuminated with the sum of the waves

$$C(x,y) = r(x,y) + o'(x,y) \qquad (15.8)$$

during the reconstruction.

The illumination of the hologram with the wave $r(x,y)$ leads to the reconstruction of the object wave $o(x,y)$, which interferes with the scattered wave $o'(x,y)$. When deriving the intensity of the interference pattern it has to be noted that a fixed phase difference of π which is given by the setup exists between the two waves o and o' even without any deformation. When calculating the amplitude transmission t which results after developing a photographic plate it is assumed that t decreases linearly with the intensity:

$$t = t_0 - \beta \tau I. \qquad (15.9)$$

Equation (15.9) is identical to Eq. (2.21) except for the sign. There the sign is contained implicit in β. This negative sign can be taken into consideration for the object wave as an additional phase term of π since $\exp(i\pi) = -1$.

15.2.3
Intensity of the Interferograms

The superposition of the undisturbed object wave o with the reference wave r leads to an amplitude transmission on the photo plate that was already calculated in Eq. (2.22a). Taking Eq. (15.9) into account this yields

$$t = \left(t_0 - \beta\tau\left[r^2 + o^2\right]\right) - \beta\tau[\mathbf{r}^*(x,y)\mathbf{o}(x,y) + \mathbf{r}(x,y)\mathbf{o}^*(x,y)]. \qquad (15.10)$$

In the following the third term which is the only one relevant will be further examined. It is the share of the virtual image t_v:

$$t_v = -\beta\tau \mathbf{r}^*(x,y)\mathbf{o}(x,y). \qquad (15.11)$$

When reconstructing the holographic image with the reference wave using Eq. (15.11) the virtual wave $\mathbf{u}_v$ becomes

$$\mathbf{u}_v = -\beta\tau r^2 o_0(x,y)e^{i\Phi(x,y)}. \qquad (15.12)$$

Using $-1 = \exp(i\pi)$ this becomes

$$\mathbf{u}_v = +\beta\tau r^2 o_0(x,y)e^{i\Phi(x,y)+\pi}. \qquad (15.13)$$

The reconstructed object wave interferes with the object wave of the deformed object o'. The deformation of the object corresponds to a phase change $\delta(x,y)$:

$$\mathbf{o'} = o(x,y)e^{i(\Phi(x,y)+\delta(x,y))}. \qquad (15.14)$$

The observed intensity in the real-time interferogram I_e is

$$I_e = (\mathbf{u}_v + \mathbf{o}')(\mathbf{u}_v + \mathbf{o}')^*. \tag{15.15}$$

Setting the amplitudes for simplicity to

$$A = \beta \tau r^2 o(x,y) \quad \text{and} \quad B = o(x,y)$$

this yields using Eqs. (15.13) and (15.14):

$$I_e = A^2 + B^2 + ABe^{i(\delta-\pi)} + ABe^{-i(\delta-\pi)}. \tag{15.16}$$

After some minor mathematical transformations this results in

$$\begin{aligned} I_e &= (A-B)^2 + 2AB(1-\cos\delta) \\ &= (A-B)^2 + 4AB\sin^2\left(\frac{\delta}{2}\right). \end{aligned} \tag{15.17}$$

The intensity is a periodical function with the same period $\delta = 2\pi$ like in the double exposure technique.

The intensity of the image becomes minimal if δ is an even multiple of 2π:

$$I_e = (A-B)^2 \quad \text{for} \quad \delta = 2m\pi, \quad m = 0, 1, 2, \ldots. \tag{15.18}$$

The intensity is maximal if δ is an uneven multiple of 2π:

$$I_e = (A+B)^2 \quad \text{for} \quad \delta = (2m+1)\pi, \quad m = 0, 1, 2, \ldots. \tag{15.19}$$

15.2.4
Visibility

A comparison of the interferometric results of the double exposure technique and the real-time holography reveals that the visibility is usually higher for the double exposure.

According to Eq. (2.37) for double exposure interferometry the optical visibility V is

$$V = 1.$$

The visibility for real-time interferometry can be calculated from Eqs. (15.18) and (15.19):

$$V = \frac{2AB}{A^2 + B^2}. \tag{15.20}$$

For $A = B$ the visibility $V = 1$. But most of the time the amplitude of the scattered wave B will be much larger than the amplitude of the undisturbed object A given by its diffraction efficiency:

$$A < B.$$

Hence the result for V when neglecting A^2 in the denominator of Eq. (15.20) is:

$$V \approx \frac{2A}{B} \ll 1.$$

To increase the visibility in real-time interferometry the wave of the scattered light B should be reduced.

15.2.5
Practical Realization

The problem in a transmission setup for real-time interferometry is the difficulty to replace the film or the photo plate to the exact same position after the developing process. The required accuracy needs to be better than $\lambda/2 = 0.3$ μm for a He–Ne laser. Only in that case the reconstructed object wave and the object wave of the undisturbed object can interfere without a phase difference. For plate holders that are very good adjustable the repositioning is accurate within a few wavelengths. However, this means that even without any deformation interference fringes are already visible. These difficulties can be overcome if the photo plate is developed at the recording position or a thermoplast film is used.

15.2.6
Thermoplast Film

The wet developing process can be omitted completely if a thermoplast film is used. The developing is then done by a heat treatment at the recording position without changing the position of the film (Section 14.4).

15.3
Fundamental Equation of Holographic Interferometry

For the evaluation of interferograms it is important to know the theoretical connection between the phase change of the scattered light wave and the geometrical displacement [34]. For the derivation of this often called "basic equation of hologram interferometry" the simplified situation in Fig. 15.2 is assumed.

In practice the displacement of an object point P to a new position P' is small compared to the distance to the light source and the hologram, respectively. According to the figure the geometrical difference of the paths of light to the hologram over point P' is:

$$\Delta s = d(\cos \varphi + \cos \varphi'). \tag{15.21}$$

From that follows a phase difference δ of

$$\delta = d(\cos\varphi + \cos\varphi')\frac{2\pi}{\lambda}. \tag{15.22}$$

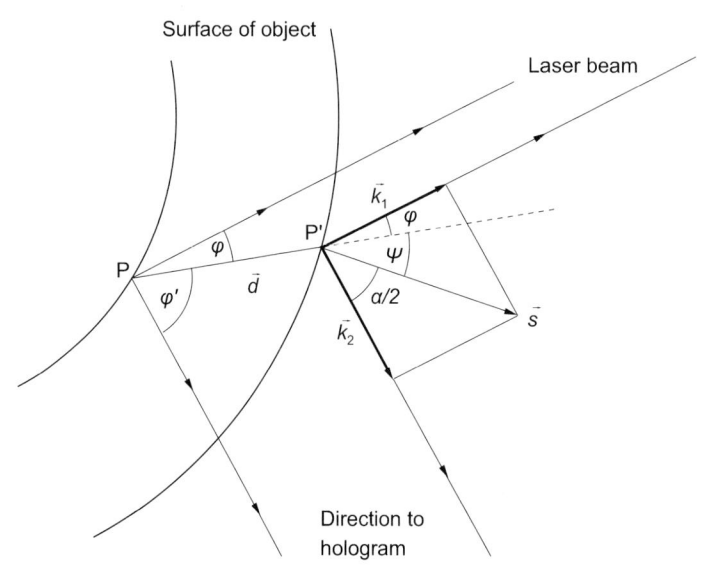

Fig. 15.2 Sketch for the derivation of the basic equation of holographic interferometry. Shown is a detail of the undisturbed object with the object point P and its position P' after a displacement of d.

This result may be written more easily using the wave vectors $\mathbf{k}_1$ and $\mathbf{k}_2$ ($k_1 = k_2 = 2\pi/\lambda$), which point into the direction of the incident and the reflected beam and the displacement vector $\mathbf{d}$ which points from the object point P of the undisturbed object to the displaced point P'. The following applies for the phase difference δ:

$$\delta = (\mathbf{k}_2 - \mathbf{k}_1)\mathbf{d}. \tag{15.23}$$

Equation (15.23) is called the basic equation of holographic interferometry. All quantitative evaluation methods to acquire the displacement vector $\mathbf{d}$ refer to this equation. (The negative sign inside the brackets of Eq. (15.23) can be avoided by orienting $\mathbf{k}_1$ and $\mathbf{k}_2$ from the object point P.) For the difference of the wave vectors the sensitivity vector $\mathbf{S}$ is defined as

$$\mathbf{S} = \mathbf{k}_2 - \mathbf{k}_1. \tag{15.24a}$$

The vector $\mathbf{S}$ lies in the direction of the bisector ($\alpha/2$) between the incident beam and the observation direction. In this direction the change in the interference field is the largest. Therefore, the vector $\mathbf{S}$ is called the sensitivity vector. According to Fig. 15.2 the length of **S** depends on the angle α between $\mathbf{k}_1$ and $\mathbf{k}_2$.

$$S = \frac{4\pi}{\lambda} \cos\left(\frac{\alpha}{2}\right). \tag{15.24b}$$

For the double exposure technique it was shown in Eq. (15.7) that the intensity is modulated with a period of 2π. If the fringe order is set to $N = 0$ at an undisturbed point of the object and the bright fringes are counted up to the object point P then $\delta = N2\pi$. In other words, if the object is continuously deformed more and more at one side then permanently new fringes appear that are counted by N. Using Eq. (15.23) yields

$$N2\pi = \mathbf{S}\mathbf{d} = Sd\cos\psi \quad \text{or}$$

$$d = \frac{N\lambda}{2\cos\psi \cos\left(\frac{\alpha}{2}\right)}. \tag{15.25}$$

Here ψ denotes the angle between the displacement vector **d** and the sensitivity vector **S**.

The different evaluation methods discussed in the literature [34] deal with two problems that make the solution of Eq. (15.25) difficult. On one hand it is often not possible to perfectly determine the fringe order N because the order $N = 0$ cannot be defined unambiguously. On the other hand, in the general case **S** and **d** are vectors with three spatial components. This problem can be solved, e.g., by recording and evaluation of three holograms that have been taken from three different directions [33, 34].

From Eq. (15.25) it can be seen that for $\psi = 0$, i.e., when the displacement vector points in the direction of the sensitivity vector, the change in the fringe order is highest. If in addition the observation direction and the illumination direction are antiparallel ($\alpha = 0$) then

$$d = \pm \frac{N\lambda}{2}.$$

The sign, i.e., the direction of the displacement, cannot be determined without great effort.

15.4
The Holo Diagram

With the development of the holo diagram Abramson [35] introduced a method that simplifies the evaluation of interferograms.

It consists, like shown in Fig. 15.3, of an arrangement of ellipses in which the holographic setup is integrated. The beam splitter and the photographic

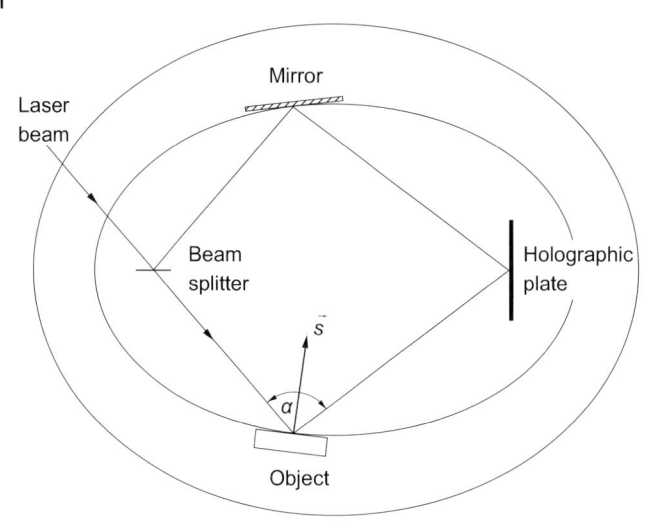

Fig. 15.3 The holo diagram by Abramson [35]. Beam splitter and photo plate are placed in the focal points of an ellipse. The path difference of reference wave and object wave measured from the beam splitter is constant if mirror and object are only moved on the ellipse (definition of the ellipse).

film are placed in the focal points. (In some setups the point light source is placed in one of the focal points).

According to its definition the sum of the distances from the two focal points is a constant for an ellipse. Therefore, the object in Fig. 15.3 can be moved on this ellipse without changing the path difference between reference and object wave measured from the beam splitter. The sensitivity vector is perpendicular to the ellipse. Moving the object in the direction of vector **S** Eq. (15.25) yields with $\psi = 0$:

$$d = \frac{N\lambda}{2\cos\left(\frac{\alpha}{2}\right)} \quad \text{or} \tag{15.26}$$

$$d = NC\frac{\lambda}{2},$$

where $C = 1/\cos(\alpha/2)$. If the object is positioned to the right of the hologram $\alpha = 0°$ can happen. This results in $C = 1$ and $d = N\lambda/2$. For all other positions the displacement d is larger than $N\lambda/2$, where N is the number of created interference fringes.

A movement of the object on the ellipse in Fig. 15.3 does not cause any path change and therefore no change in the system of interference fringes. This effect is used to create usable holograms on tables with low stability.

The holo diagram can also be useful in other holographic problems. Creating two ellipses with the same focal points (Fig. 15.3) such that the sum of the distances from the focal points to one point of the outer ellipse is larger by the coherence length than the space between the ellipses marks the area which an object can fill without exceeding the coherence length.

15.5
Time Average Interferometry

15.5.1
Theory

The time average interferometry is used to measure the oscillation amplitude of an object oscillating with a frequency of $f = \omega/2\pi$. Here the exposure time is much larger than the oscillation period $2\pi/\omega$. The displacement vector can be written as

$$\mathbf{d} = \mathbf{d}(r)\sin(\omega t).$$

For simplification we assume an oscillation in the z-direction, which is shown in Fig. 15.4 for the two-dimensional case. For harmonic oscillations applies:

$$\mathbf{d} = \mathbf{d}(z)\sin(\omega t).$$

The according phase shift δ is

$$\delta = \mathbf{S}\mathbf{d}\sin(\omega t).$$

For the complex amplitude of the object wave this results in

$$\mathbf{o}(z,t) = o(z)e^{i(-\Phi + \mathbf{S}\mathbf{d}\sin(\omega t))}. \tag{15.27a}$$

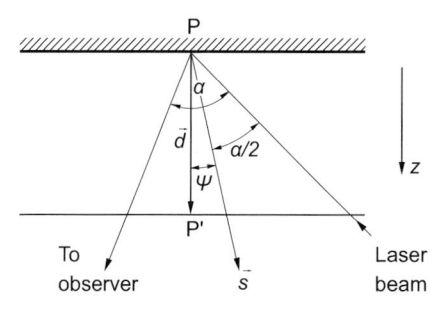

Fig. 15.4 Time average interferometry in a two-dimensional example. A sinusoidal oscillation causes a periodic displacement of the object point P into the new position P'. **S**: sensitivity vector, **d**: displacement vector.

The reconstructed wave $\mathbf{u}(\mathbf{z})$ is the time average of many individual waves registered during the exposure time τ according to Eq. (15.27a):

$$\mathbf{u}(z) = \frac{1}{T}\int_0^T \mathbf{o}(z,T)dt \quad \text{or}$$

$$\mathbf{u}(z) = o(z)e^{-i\Phi}\frac{1}{T}\int_0^T e^{i\mathbf{Sd}\sin(\omega t)}dt. \tag{15.27b}$$

The following expression

$$M = \frac{1}{T}\int_0^T e^{i\mathbf{Sd}\sin(\omega t)}dt \tag{15.28}$$

is called the "characteristic function" or "modulation function." The integral can be expanded into a power series of Bessel functions. Since the exposure time is very long only the Bessel function J_0 adds to the integral:

$$M = J_0(\mathbf{Sd}). \tag{15.29}$$

For Eq. (15.27b) follows:

$$\mathbf{u}(z) = o(z)e^{-i\Phi}J_0(\mathbf{Sd}).$$

The observed intensity $I \propto |\mathbf{u}|^2$ is then

$$I(z) = I_0 J_0^2(\mathbf{Sd}). \tag{15.30}$$

Figure 15.5 shows the slope of J_0^2; the intensity maxima decrease very quickly with increasing order N.

The determination of $\delta(=\mathbf{Sd})$ from the order N is a little bit more complicated than for the interference techniques mentioned before because although J_0^2 is an oscillating function it is not periodic. If the fringe order is not an integer or half-integer number the needed interpolation methods get quite complicated. However, positions of zeroth order, i.e., nodal points, can be found very easy with this method since they are reconstructed with full intensity ($J_0^2 = 1$); Fig. 15.5.

15.5.2
Practical Realization

An often realized experiment is the measurement of oscillations of a loudspeaker. To realize the shown relationships it is driven with only one frequency since a harmonic oscillation was assumed in the theoretical analysis. It is best to use a frequency generator to create interferograms at different frequencies. The frequency generator is positioned beneath the holographic table to avoid disturbing vibrations. It has to be made sure that the vibrations of the object do not affect the complete setup.

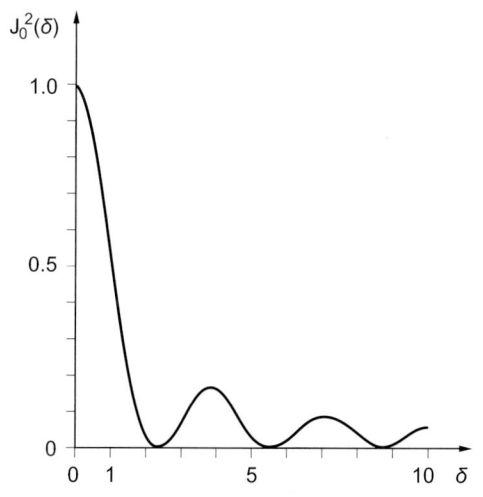

Fig. 15.5 The square of the Bessel function $J_0(\delta)$. The (not equidistant) values of $\delta = Sd$ for the maxima and minima can be seen in the figure.

To enhance the contrast a so-called stroboscopic method was proposed in which a triggered double exposure always records only two exact oscillation conditions. This results in conditions that were discussed for the double exposure method and the contrast is maximal for all orders.

15.6 Speckle Interferometry

The backscattered light of a coherent illuminated object shows a grainy structure which is called "granulation" or "speckles." The reason for this is the roughness of the surface which acts like a statistical reflection diffraction grating. In many holographic applications this effect is disturbing. On the other hand, it can be thought of as an interference field of the statistically distributed surface roughness of the object. Hence this statistical distribution of bright and dark spots carries information about the condition of the objects surface.

The granulation can be used for the interferometrical investigation of deformations. In "speckle photography" a coherent light source is needed, however there is no need for a reference wave. The object is illuminated with a laser wave and the scattered light is imaged onto a film with fine granulation using a lens. Since the speckle pattern is comparatively coarse no holographic film material is needed. A second recording of the then slightly deformed object is superposed with the first one. The two, for the eye not distinguishable, statistical distributions of the speckles superpose on the film and create interference fringes. From these the amount of deformation of the object can be determined. The sensitivity of this method is given by the size of the speckles and is about an order of magnitude lower than for the other methods described in this chapter [34].

In "speckle interferometry" an object wave and a reference wave is used like in other interferometric setups. A simple setup consists of a Michelson interferometer (Fig. 8.10) in which one of the mirrors is replaced by the diffuse reflecting surface of the object. If the surface is imaged onto a screen which is placed in Fig. 8.10 at the position of the detector using a lens then the image is covered with speckles, which were created by the interference of object and reference waves. Every small change of the position of the scattering surface in the beam direction leads to a change in the observed speckle field. For slow movements a flicker in the image will be observable. For an oscillating membrane the speckles disappear in the moving areas while the nodal points still show speckles. For applications and further developments in the area of speckle interferometry the reader is referred to the literature [36].

Problems

Problem 15.1 Show that for double exposure interferometry the real image shows the same interference pattern than the virtual image (Eq. (15.7)).

Problem 15.2 Calculate the visibility of the interference fringes for double exposure interferometry.

Problem 15.3 Verify Eq. (15.17) from Eq. (15.16).

Problem 15.4 Verify Eq. (15.20) from Eqs. (15.18) and (15.19).

Problem 15.5 Observing an interferogram between two points there shall be eight dark interference fringes. It should be $\psi = 0$. And the angle between illumination wave of the object (k_1) and the observation direction (k_2) shall be 90°. Calculate the deformation ($\lambda = 632.8$ nm).

Problem 15.6 For $a = 0$ and $\psi = 0$ the value of d is $d = N(\lambda/2)$. Why is the deformation a multiple of $\lambda/2$ and not of λ?

Problem 15.7 A time average experiment of a loud speaker shows a bright spot ($N = 0$) in the center and in some distance another bright concentrically ring. Between there is a dark ring. What is the deformation amplitude d of the dark ring compared to the surface in the middle of the (unusual) flat loud speaker? ($\lambda = 632.8$ nm, Fig. 15.4, $\alpha = \Psi = 0$)

16
Holographic Optical Elements

To understand holograms it is important to note that they have the same effect as a Fresnel zone plate and can be thought of as complicated gratings. From the first property follows their ability to reconstruct images which can be compared to the imaging process of lenses. The second aspect explains the chromatic aberration during reconstruction.

These two characteristics are the basis for the manufacturing of optical components by means of holography. The advantages compared to the conventional optical elements (e.g., glass lenses) are small thickness of the elements and the possibility to develop optical elements that cannot be created in a conventional way.

16.1
Lenses, Mirrors, and Gratings

16.1.1
Lenses and Mirrors

The production of holographic lenses and mirrors does not require additional knowledge of holography and its application areas [27]. Figure 16.1 shows the effect of a diverging lens. Rays parallel to the axis are refracted when passing the lens. The divergent rays seem to come from the virtual focal point F.

The same effect can be obtained by a transmission hologram that was created from a spherical object wave and a plane reference wave. The illumination of the hologram with a plane wave that resembles the reference wave creates the diverging object wave, apparently coming from the focal point F. When using a conjugated reference wave (shown as a dashed line in Fig. 16.1) the real image is reconstructed, i.e., the rays are focused on the real focal point F'. Like every zone plate the holographic lens shows the properties of a diverging and also of a collecting lens (Fig. 2.9).

The above statements can also be applied to the imaging by a concave mirror. Figure 16.2 shows the ray paths for such a mirror. And again the hologram combines properties that in optics can only be realized by two different com-

16 Holographic Optical Elements

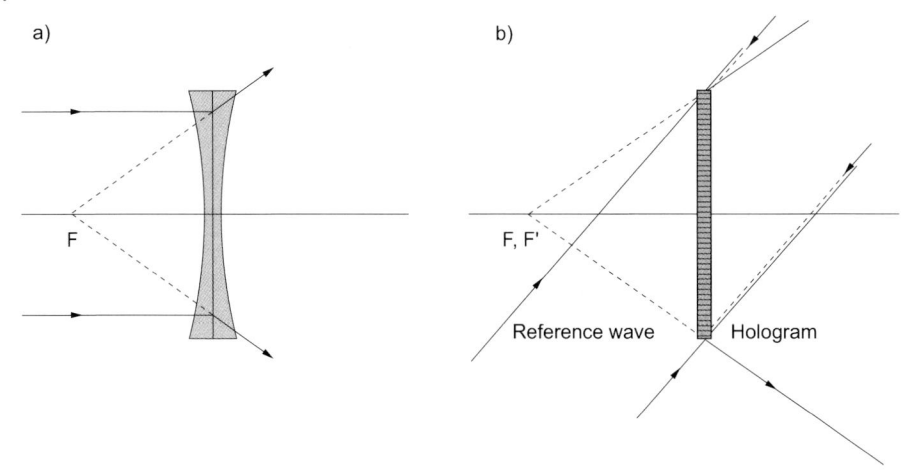

Fig. 16.1 Comparison of (a) a diverging lens and (b) a transmission hologram that was created by a plane reference and a spherical object wave. When reconstructing with a plane reference wave the hologram has the effect of a diverging lens; using a conjugated reference wave shows the effect of a collecting lens.

ponents. The reconstruction using the reference wave creates the diverging original wave, emerging from F', which was used for the production of the reflection hologram. The hologram acts like a convex mirror. When working with the conjugated reference wave (shown as a dashed line in Fig. 16.2) the conjugated object wave is created. This is a convergent wave with the real focal point F. The hologram now has the property of a concave mirror. The reflection hologram is both a concave and a convex mirror. The production of lenses and concave mirrors according to Figs. 16.1 and 16.2 is also possible using an axis parallel reference wave. In the case of a parallel reference wave the virtual and real focal points F, F' coincide. If a divergent or convergent reference wave is used F and F' are different. This will be shown in the next section.

16.1.2
Focal Length

Holographic lenses can be produced according to Fig. 16.3 using a divergent object wave and a parallel or divergent reference wave. In Fig. 16.3 both waves are divergent from point like coherent light sources and the distances from the sources and the holographic film are s_{obj} and s_{ref}. Both values are positive in Fig. 16.3. If a wave is incident from the other side on the holographic film the distance is defined negative. The lens in Fig. 16.3 is a diverging lens with the

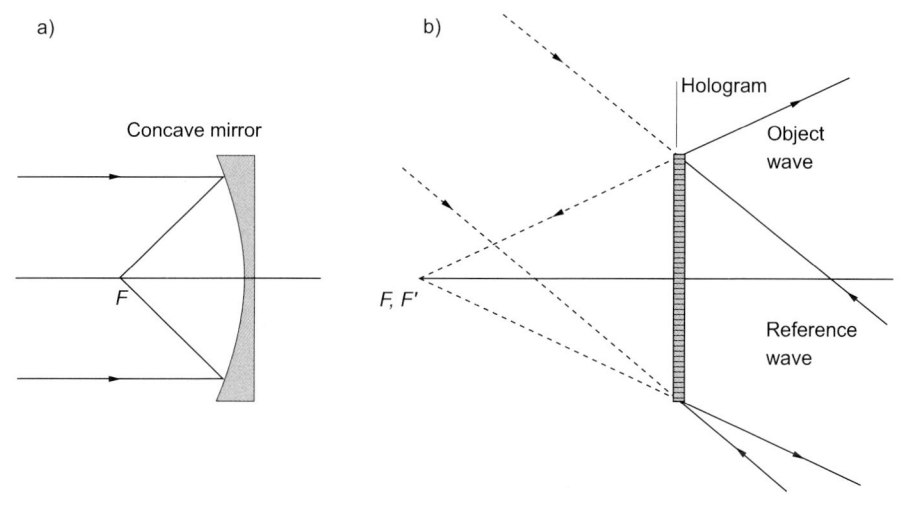

Fig. 16.2 Comparison of (a) a concave mirror and (b) a reflection hologram that was created from a plane reference wave and a spherical object wave. When reconstructing with the reference wave the hologram acts as a convex mirror, when reconstructing with the conjugated reference wave it acts as a concave mirror.

focal length f which can be calculated using the equation

$$\frac{1}{s_{obj}} + \frac{1}{s_{ref}} = \frac{1}{f} \qquad (16.1)$$

where f is the focal length of the *diverging lens*.

It is possible to use the *diverging lens* of Fig. 16.3 also as a *collecting lens*. In this case the reference wave has to fall from the opposite direction on the hologram. This is not the conjugated wave because the curvature remains unchanged. For the calculation of the focal length f' of the *collecting lens* the distance s_{ref} is used with a negative sign in (16.1):

$$\frac{1}{s_{ref}} - \frac{1}{s_{obj}} = \frac{1}{f'}. \qquad (16.2)$$

According to Eqs. (16.1) and (16.2) the focal length of the diverging and collecting lens f and f' are different. Only in the case of a parallel reference beam with $s_{ref} = \infty$ we have $f = f'$. This case is shown in Figs. 16.1 and 16.2. In the coordinate system used in this section f and f' are both positive.

Figure 16.3 shows a setup for the production of a holographic lens for angled light incidence. It is also possible to use the same axis for the object and reference waves. As an example in Fig. 16.3 it was assumed that $s_{obj} = (1/2)s_{ref}$. From Eqs. (16.1) and (16.2) it follows that the focal length of the diverging and collecting lens are $f = s_{ref}/3$ and $f' = s_{ref}$, respectively.

16 Holographic Optical Elements

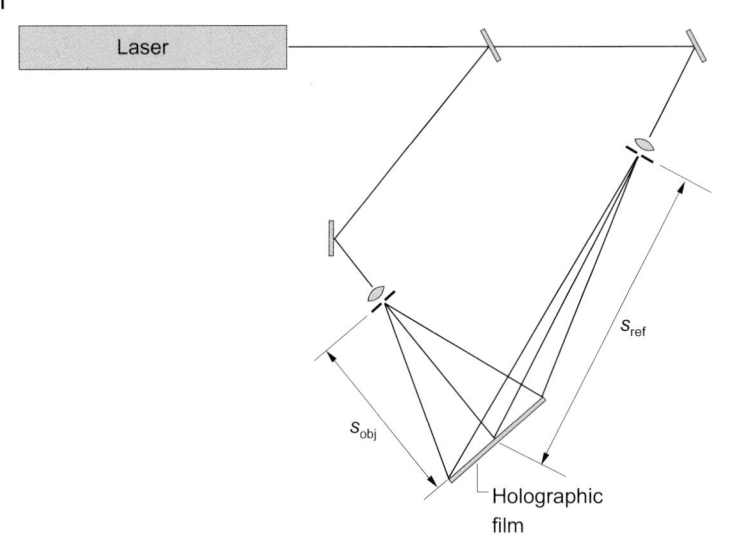

Fig. 16.3 Setup for the production of a holographic lens (for angled light incidence). For the creation of the transmission hologram two spherical waves are superposed. As an example the distance s_{ref} is twice as large as the distance s_{obj}. According to Eq. (16.2) $f' = s_{ref}$.

To create a holographic mirror a reflection setup is used. The selection of distances and optical elements is similar to the setup for a lens. The setup shown in Fig. 16.4 uses an angle of 45° for the reference wave similar to Fig. 16.3. As an example the length s_{ref} is twice as large as the distance s_{obj}. From that it follows that $f' = s_{ref}$. The production of lenses and mirrors can also be done with a plane reference wave. In Fig. 16.4 s_{obj} and s_{ref} are defined both positive. If the direction of incidence is changed a negative sign has to be applied.

Figure 16.5 illustrates the relation between f', s_{ref}, and s_{obj} for lenses and mirrors according to Eq. (16.2). Here s_{ref} and s_{obj} are the distances of the reference and the object point light source. For s_{ref}/s_{obj} and $f' = s_{ref}$ it follows that $f'/s_{obj} = 2$. The larger s_{ref} gets compared to s_{obj} the closer f' gets to the value of s_{obj}. In the case of a mirror for $s_{ref} = s_{obj}$ a plane mirror is created.

A holographic lens corresponds to a Fresnel zone plate. From Eq. (2.36) it becomes clear that the imaging by a holographic lens comes with a strong chromatic aberration. For $z_o = f$ and $k^2 \lambda^2 \ll 2 z_o k \lambda$ it follows that

$$f = \frac{r_k^2}{2k\lambda} \sim \frac{1}{\lambda} \vartheta \quad \text{or} \quad \frac{f_{\lambda 1}}{f_{\lambda 2}} = \frac{\lambda_2}{\lambda_1}. \tag{16.3}$$

The focal length f therefore decreases with increasing wavelength. The symbols in Eq. (16.3) are explained in Fig. 2.8.

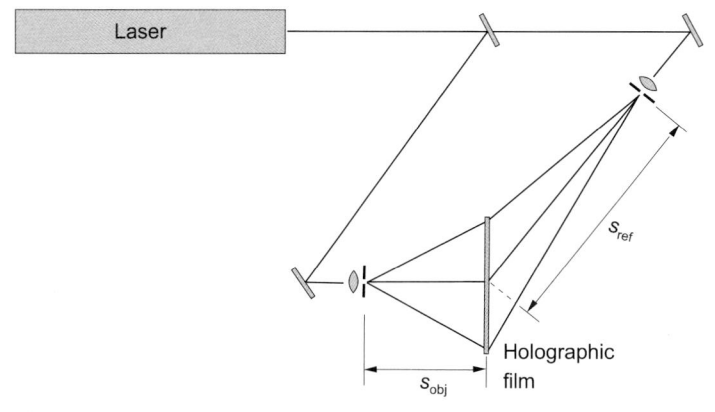

Fig. 16.4 Setup for the production of a holographic mirror with angled light incidence. The denotations correspond to that of Fig. 16.3. The HOE represents a reflection hologram.

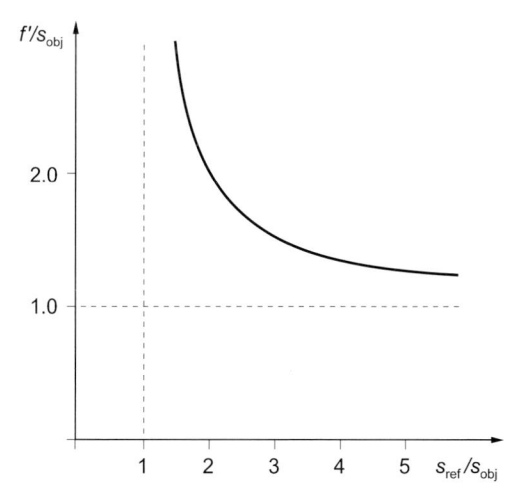

Fig. 16.5 Diagram of Eq. (16.2). The connection of the normalized parameters f'/s_{obj} and s_{ref}/s_{obj} can simplify the design of HOEs.

16.1.3
Gratings

The production of a simple grating is described in Section 8.5. For gratings with spatial frequencies of 1000 mm^{-1} and more a transmission setup can be used (Chapter 10). The two used waves are plane. (The denotation reference and object wave is now superfluous.) The smaller is the grating constant the larger is the angle between the plane wave. For special applications the properties of lenses and gratings can be mixed by choosing between plane and spherical waves.

A so-produced sinusoidal grating should only reconstruct one diffraction order. However, most often observers see more than one order. As explained in Chapter 6 a pure volume sinusoidal grating is the exception. Especially in the transition area from thin to volume gratings one cannot expect only one diffraction order. Very often the result is also influenced by the developing process because for high contrast also the nonlinearities of the H&D curve have to be considered. The grating recorded in the emulsion is then not a pure sinusoidal grating anymore but a mixture between a line and a sinusoidal grating. Higher diffraction orders caused by the H&D curve can be suppressed by a lowered contrast.

16.1.4
Beam Splitters

The diffraction efficiency of hologram often lies well below 100%. In addition to the first diffraction orders which correspond to the virtual and the real image there exists always an intense zeroth order.

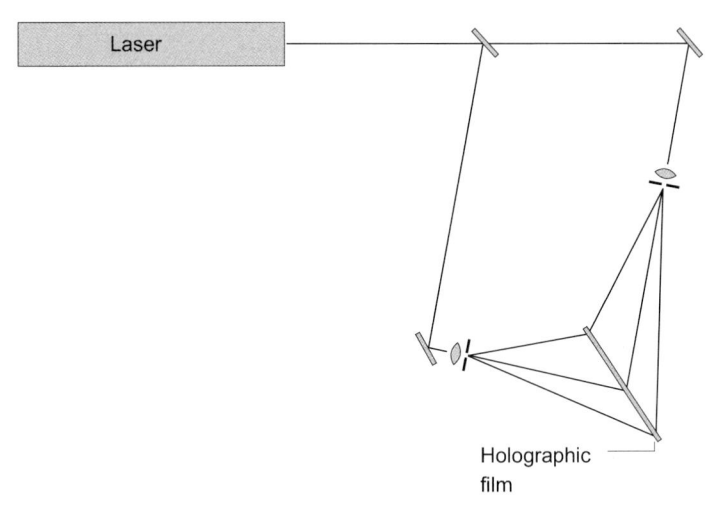

Fig. 16.6 Setup for the production of a beam splitter. The reflection hologram is created by two spherical waves that are perpendicular to each other. If $s_{ref} = s_{obj}$ a plane mirror is created that can be used as a beam splitter because its diffraction efficiency is less than 1.

Therefore, holograms can also be thought of as beam splitters. For the production of a beam splitter the setup shown in Fig. 16.6 can be used. In this figure the two spherical waves are perpendicular to each other. The angle of the reference wave is chosen in a way that the directly reflected wave does not hamper the beam splitting. If the reconstruction does not occur under the Bragg angle the beam splitting is shifted toward the transmitted wave. Also the intensity distribution can be controlled this way. Such a beam splitter cannot be used far way from the Bragg angle since its diffraction efficiency then drops to zero.

16.2
Computer-Generated Holograms

16.2.1
Complex HOEs

In addition to the simple holographic optical elements holography offers the possibility to produce complicated HOEs that cannot be produced by simple means in the area of glass optics. Technical applications are, e.g., in the area of optoelectronics in the production of beam splitters with more than two splits or in the production of lenses with more than one focal length [27].

A totally different procedure to produce HOEs is given by the use of computers [37,38]. Computer-generated holograms (CGHs) offer the possibility to calculate HOEs without the use of a laser, to print them and to convert them into real holograms by using photolithographic processes. Diffractive optical elements (DOE) find an increasing application, e.g., in laser technology, in optoelectronics and in the information technology.

16.2.2
Calculated HOEs

It is not within the scope of this book to give an in-depth presentation of the theory and the production of CGHs. The developments in computer technology have increased the importance of this field of holography and today it represents its own discipline [37]. With the computer it is even possible to produce holograms of objects that do not exist or whose experimental setup is not feasible. So CGHs are applied in complicated technical applications. A more detailed view is given in Chapter 19.

To produce a CGH the complex amplitude of the object wave in the hologram plane needs to be calculated. This is the Fourier transform of the amplitude in the object space. The object wave is calculated pixel wise and it has to be made sure that the number of object points is large enough. A good image can be achieved with a field of 1000×1000 pixels. The computing time for the discrete Fourier transform is extremely large due to the many complex multiplications in two dimensions. For the mentioned field size 10^{12} operations are required. The development of computer holography is therefore characterized by the search for methods that shorten the computing time. One way to achieve this is the use of the fast Fourier transform (FFT). After calculating the amplitude of the object wave the reference wave is added and the intensity is calculated.

For the production of the CGH the calculated intensity distribution is transferred with high precision onto a suitable medium. By using an electron beam writer a mask four to ten times larger than the resulting CGH can be created. This mask is then imaged scaled down on a wafer that is coated with photosensitive lacquer. Here a silicon wafer can be used that might be coated with a SiO$_2$ layer of 1 μm thickness. After exposure the exposed (or the unexposed) areas of the photo lacquer are dissolved so that the surface of the wafer is partly exposed. By a dry etch procedure the wafer surface is etched in these areas. The etch depth is 1/6 of the wavelength meaning 0.14 μm for a wavelength of 0.84 μm. The result is a surface relief that represents the CGH as a phase hologram. The hologram can be duplicated with the help of embossing techniques (Section 17.1).

The CGH produced by this method using only a single mask is a binary hologram, i.e., the holographic surface relief has a step profile with only two planes. While sinusoidal structures only have one diffraction order a binary step profile exhibits several diffraction orders. Therefore, a step like approximation of a sinusoidal profile is desired for CGHs (Fig. 16.7). Already a structure with three steps suppresses the higher orders sufficiently. However, three masks and three etching steps are needed for the production. During this the masks have to be positioned with high accuracy for the exposure, i.e., in fractions of a wavelength.

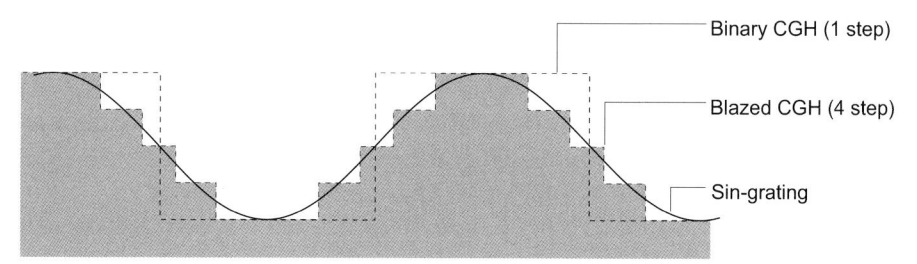

Fig. 16.7 Surface structure of a sinus grating, binary CGH and a CGH with four masks.

CGHs can also be transferred onto normal holographic emulsion layers. To do so the calculated hologram is transferred to a plotter, photographed, and imaged scaled down onto the holographic layer. Depending on the emulsion type an amplitude or a phase hologram can be created as a CGH this way.

16.3
Electronic Holography

Digital holography includes CGHs, electronic holographic displays, and image detection. CGHs were discussed in the last section and electronic holography will be explained in the following one.

16.3.1
Angle of Diffraction in Electronic Holography

In electronic holography a liquid crystal display or a digital micromirror device (DMD) can be used as a holographic medium. The pixel size of liquid crystal or other displays is about 5 µm or larger. Thus the lowest holographic grating constant for such a display is about $d_g = 10$ µm or more. According to Fig. 2.7 a holographic grating is created by the superposition of a reference and an object wave with an angle $\alpha + \beta$ between them. Using Eq. (2.35) for the first order of diffraction ($N = 1$)

$$d_g = \frac{\lambda}{\sin\alpha + \sin\beta} \tag{2.35}$$

resulting for $\lambda = 0.6$ µm in an angle $\alpha + \beta \approx 0.06$ rad $\approx 3.4°$. For reconstruction the angle of diffraction has the same value. We may conclude that a display with a pixel size of 5 µm can produce a diffraction angle between 0° and 3.4° depending on the spatial frequency or grating constant. A CGH written on the LCD display consists of a superposition of gratings with $d_g \geq 10$ µm. In about 1 m distance this yields a hologram with a maximum size of about 0.06×1 m = 6 cm. A similar calculation can be performed using the equation for the Fresnel lens in Section 2.6. Due to the pixel size of available displays electronic holography is limited to diffraction angles of several degrees which is sufficient for many applications.

16.3.2
Holographic Electronic Display

A typical holographic display consists of a liquid crystal display (LCD) as a spatial light modulator. Polarization filters have to be removed from the LCD. The pixel size is 5, 8, or 16 µm with about 1920 × 1080 pixel. The fill factor is over 90% and the dynamic range is given by 8 or 12 bits. Liquid crystals show birefringence which can be controlled electronically in every pixel. This results in a change of the refraction index and a phase shift of a polarized laser beam passing the area of the pixel. Applying a voltage the phase shift can be changed between 0 and 2π and a phase grating can be electronically written on the display. The diffraction efficiency is about 90%. Computer-generated holograms can be displayed with video frequencies. The LCD display can operate in transmission or in reflection using a mirror behind the display as shown in Fig. 16.8.

It is possible to detect an optical hologram with a digital camera and transfer the electronic signals via a PC to the LCD display. A setup for recording and optical reconstruction of digital holograms is shown in Fig. 16.8. A laser (e.g., a frequency doubled 150 mW Nd:YAG laser at 532 nm) is split into two beams (reference and object wave) creating a hologram on a digital camera (without

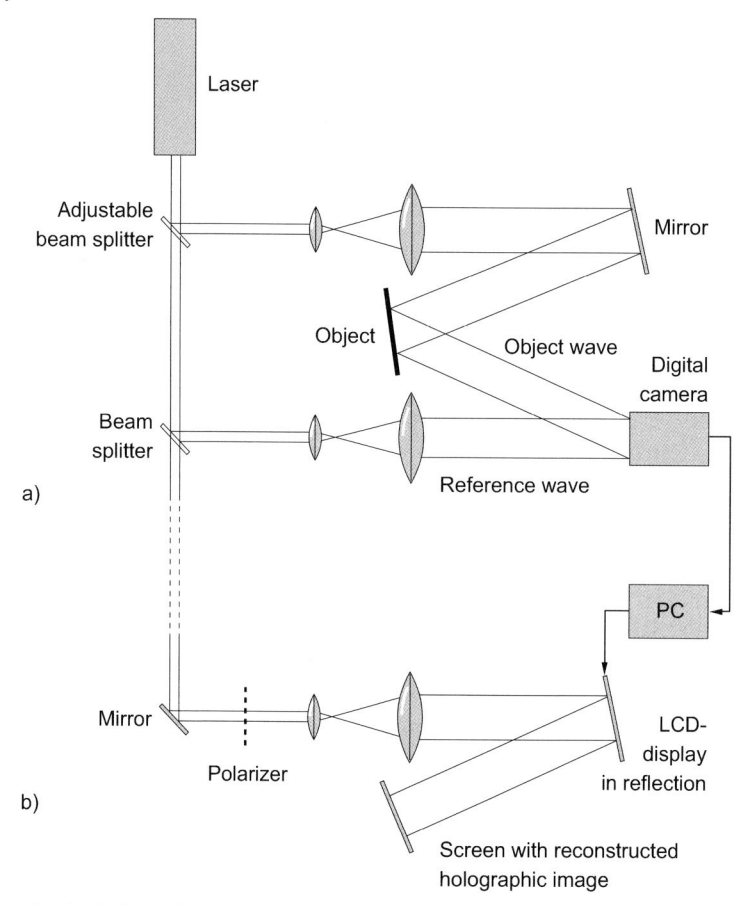

Fig. 16.8 Setup for electro-optical recording and reconstruction of digital holograms: (a) an optical hologram from an object is detected by a digital camera and send to a PC and (b) the PC sends signals to the LCD display creating a hologram. The reconstruction of the image is performed by a laser beam. LCD displays can operate in transmission by removing the mirror behind the display.

objective) (Fig. 16.8). The data from the camera are transferred to a PC which gives the electronic signals to the display. A third beam reconstructs the image from the LCD display (Fig. 16.8b). It is also possible to give CGHs from the PC directly to the display. In this case the arrangement of Fig. 16.8a is not necessary.

The setup of Fig. 16.8 can be used for holographic interferometry for measuring small deformations or vibrations. In this case interference fringes are produced in the PC by superposition of two holograms from an object with different deformations. The object resolution is about 5 μm. The arrangement of Fig. 16.8b can be used as a 3D display of CGHs.

Problems

Problem 16.1 Describe and calculate the construction of a diverging lens with a focal length of $f = 20$ cm?

Problem 16.2 Describe and calculate the construction of a collection lens with a focal length of $f' = 20$ cm?

Problem 16.3 Describe and calculate the construction of a diverging mirror with a focal length of $f = 20$ cm?

Problem 16.4 Describe an calculate the construction of a collecting mirror with a focal length of $f' = 20$ cm?

Problem 16.5 How large are f and f' for a holographic lens which is produced according to Fig. 16.3 with $s_{ref} = 30$ cm and $s_{obj} = 10$ cm? Explain the meaning of f and f'.

Problem 16.6 A holographic grating is produced by two plane waves from a He–Ne laser ($\lambda = 632$ nm)? The angles of incidence are $0°$ and $45°$. Calculate the grating constant d_g and the spatial frequency σ. Describe the condition for a transmission and reflection grating.

Problem 16.7 How large is the image of an electronic hologram using a LCD display with a pixel size of 500 nm?

17
Security and Packing

17.1
Embossed Holograms

It is not possible to print holograms on paper by the conventional printing technology. The resolution of the printing plates and of paper is two orders of magnitude lower than the grating structure of holograms in the submicrometer region. In addition, the diffraction efficiency of amplitude holograms printed on paper can be estimated to be of the order of about 3%. Thus, a special printing process to print holograms was developed which is known as embossing. It employs a technology closely related to the production of CDs and DVDs.

Embossed holograms became by far the cheapest way for mass production in holography. They are applied mainly in security applications against counterfeiting, in packing technology, and advertising. Many holographic companies exist, manufacturing and promoting these products. An embossed hologram consists of a holographic surface relief in an aluminized plastic foil. It represents a phase hologram. In the following, the production of embossed holograms is described. It consists of several steps: production of the master hologram in photoresist (Figs. 17.1 and 17.2), production of the embossing stamper or shim (Fig. 17.3), embossing of the holograms (Fig. 17.4), and hot-stamping process on product (Fig. 17.5) [39].

17.1.1
Production of the Master (Figs. 17.1 and 17.2)

For creating a surface relief in photoresist, the interference fringes of the holographic grating structure have to be directed more or less perpendicular to the surface (Fig. 17.2a). This is the case if the object and reference waves are incident to the holographic film from the same side. Thus, transmission holograms are suitable for the master. An example for the experimental arrangement for producing the master in photoresist is shown in Fig. 17.1. The real

Holography: A Practical Approach. Gerhard K. Ackermann and Jürgen Eichler
Copyright © 2007 WILEY-VCH Verlag GmbH & Co. KGaA, Weinheim
ISBN: 978-3-527-40663-0

17 Security and Packing

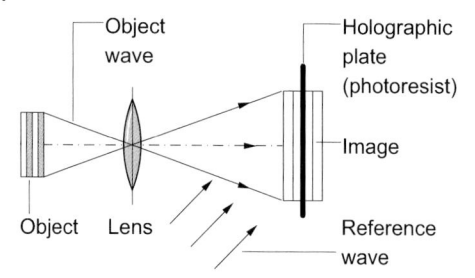

Fig. 17.1 Experimental arrangement for producing a master hologram in photoresist for embossing. Instead of using an object and a lens, a hologram may be used and the photoresist is placed into the real image of this hologram.

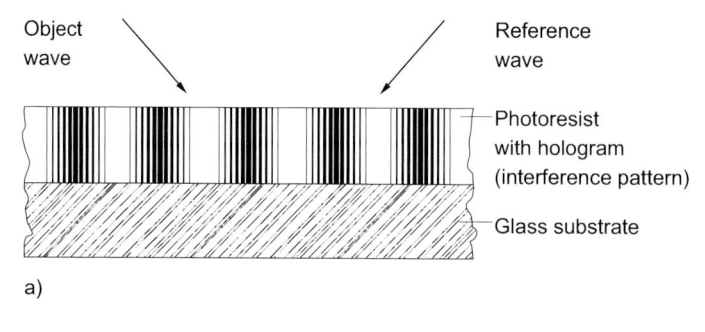

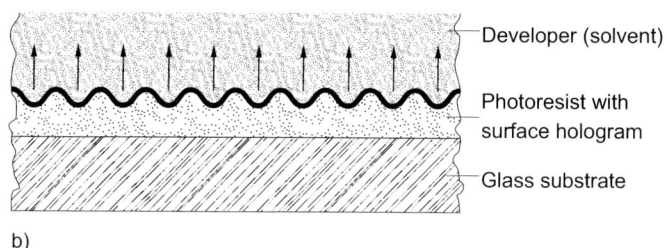

Fig. 17.2 Production of the master hologram in photoresist.
(a) Exposure by superposition of an object and reference wave.
(b) Creating a suface relief by a solving developer.

image of an object is focused by a lens on the holographic plate. The same laser is used to illuminate the object and to create the reference wave. Instead of using the image of an object, the real image of another hologram may be used. In this case, the lens in Fig. 17.1 is not necessary. Frequently rainbow transmission holograms are used, which yield impressive optical effects. In a later step, the surface structure is embossed on an aluminized thermoplastic film and the aluminum layer reflects the incident light resulting in some kind of reflection hologram.

17.1 Embossed Holograms

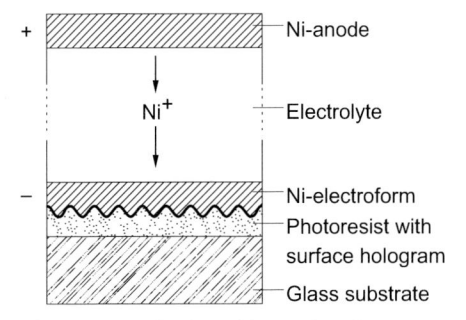

Fig. 17.3 Production of the embossing stamper or shim by nickel electroforming.

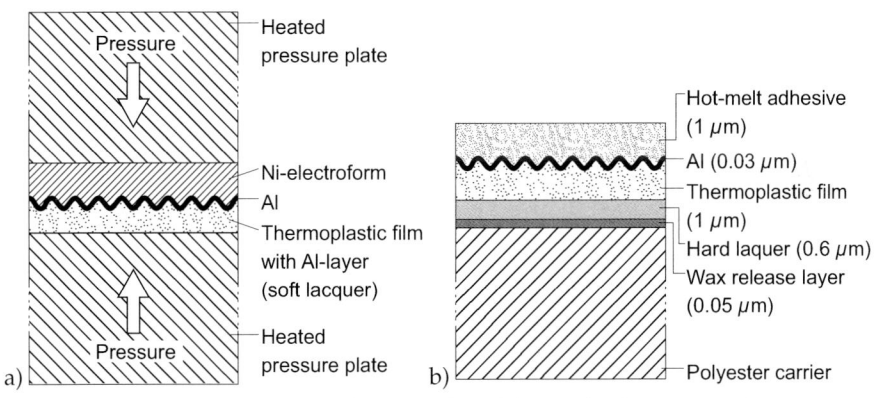

Fig. 17.4 (a) Embossing a hologram in an aluminized thermoplastic foil. (b) Several layers are added to the hologram: a hot-melt adhesive for fixing the hologram on the product (Fig. 17.5), a hard laquer to protect the hologram on the product (Fig. 17.5), and a wax release layer to connect the hologram with the polyester carrier.

The master hologram is produced in a photoresist film. A photoresist is a transparent material that becomes insoluble (negative photoresist) or soluble (positive photoresist) on exposure of blue light (Section 14.4). Some holographic groups use Shipley photoresist as a standard for embossing masters. It can be obtained on glass plates commercially and has a long shelf time of several months or years. The resolution is about 1500 lines/mm or more. This material is also used for diffractive optical elements and it is easily metalized for the production of embossing shims. It is also possible to coat glass plates with liquid photoresist in the lab using a spin table. Usually, the blue 488 nm line of an argon laser with a power of several watts is used with a power density of about 0.1 to 1 J cm^{-2}. Thus, relatively long exposure times are necessary and vibrations of the holographic arrangement have to be avoided. Photoresist is not sensitive to red and less sensitive to green light.

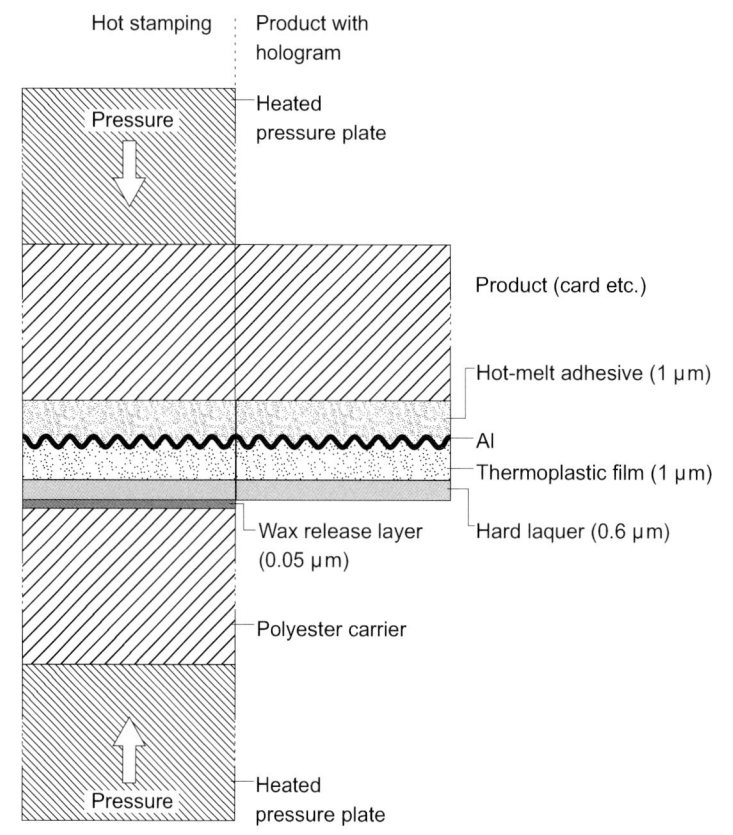

Fig. 17.5 Left side: hot stamping of the embossed hologram on the product. Right side: product with the hologram.

After the exposure, the plate is placed in a tray of developer and agitated constantly for about 10 s (Fig. 17.2b). For a positive, photoresist material is solved by the developer at the surface on the exposed regions. Within certain limits, the removal rate is proportional to the exposure. For example, the Shipley 3003A developer can be used mixed with five parts distilled or de-ionized water to one part developer. The plate is removed and rinsed for about 2 min with de-ionized water using a spray nozzle to stop developing. The plate has to be dried quickly. To avoid water spots, filtered compressed air or nitrogen can be used. The developed plates are still sensitive to light and they should be stored in light proof boxes. Inspection under daylight should be kept to a minimum. The structure of surface of the developed photoresist is shown in Fig. 17.2b. It represents a phase transmission hologram. The depth variation is about 1 μm and the fringe spacing about 0.7 μm.

17.1.2
Production of the Shim (Fig. 17.3)

The first stage in producing the embossing stamp is to make the surface of the photoresist master electrically conductive. A thin layer of metal is deposited chemically or by vacuum deposition. Figure 17.3 shows the production of a nickel electroform that is grown by standard electrolytical methods using a nickel anode and an electrolyte. A voltage is applied between the conductive photoresist surface and the nickel anode. An electroform is grown and carefully stripped from the surface and mounted on a pressure plate or a roller.

17.1.3
Embossing of Holograms (Fig. 17.4)

The pressure plate with the electroform is heated and the holograms are embossed in an aluminized thermoplastic material (Fig. 17.4a). Some groups use a thermoplastic material without aluminum and the aluminization is performed after embossing. For mass production, a process with a heated roller is more economical yielding a throughput of 30 m/min or more for a material with a width from 0.2 to 1 m. The backside of the embossed hologram is coated with self-adhesive (wax release layer) and a protective polyester carrier or paper (Fig. 17.4b). The front side is coated with a hot-melt adhesive for fixing the hologram on the product, e.g., a credit card.

17.1.4
Hot Stamping (Fig. 17.5)

In the final step, the hologram is hot stamped on the product (Fig. 17.5, left side). After stamping, the polyester carrier and the wax release layer are removed (Fig. 17.5, right side). The incident light is reflected from the aluminized holographic structure and the transmission hologram is now seen in reflection. Comparing Figs. 17.2 and 17.5, it can be seen that reconstruction of the phase hologram is made from the back side. This fact does not alter the image. It is also completely unimportant if a positive or negative photoresist is used.

17.1.5
Properties of Embossed Holograms

A drawback of embossed holograms is that they do not have a lot of depth. A typical value for embossed 3D holograms is about 2 to 3 cm (1 inch). Frequently 2D/3D holograms are produced, which are made using several 2D objects in different planes. This is shown in Fig. 17.1 where the 3D objects consist of different 2D objects. These objects may be computer-generated images on a liquid crystal display. The photoresist is exposed successively by each of these images. Thus by this type of multiplexing three-dimensional effects for the viewer can be produced.

17.1.6
Dot Matrix Hologram

A dot matrix hologram consists of an array of smaller units or dots embossed in an aluminized thermoplastic layer. Each dot is a separate microhologram which is usually a tiny diffraction grating or holopixel with different orientations on the surface. When illuminated with white light, each micrograting produces a spectrum which is reflected into a well-defined direction. The gratings can be arranged to produce an image or a variety of special effects. Dot matrix holograms are computer generated and the angle, size, shape, and spacing of each holopixel can be controlled. Dot matrix holograms are extremely eye-catching and bright. These computer-generated holograms are applied for security and decorative packaging.

17.1.7
Applications

Embossed holograms are applied frequently on credit cards, banknotes, passports, and other documents, and as labels to protect products against counterfeiting. They have also appeared on postage stamps, in book illustrations, and decorative elements on products. They can be increasingly found on gift wrappings, other packing products, stickers, and the souvenir market.

17.2
Holographic Security Devices in Industry (Counterfeiting)

The production of embossed holograms is technically a complicated process. Thus, these products are applied widely against counterfeiting of documents and products. They are well-established in many industries and are found on hosts of products and packaging, e.g., CDs and DVDs, computer software, watches, and cosmetics. Other uses are clothing hang tags, documents, certificates for fine jewelry, and others, tickets, passes, identification and credit cards. Diffractive optical elements such as DOVIDs (diffractive variable image device) protect passports, driver's licenses, bonds, and documents. However, holograms can be counterfeited by several processes and special countermeasures are necessary [40, 41].

17.2.1
Counterfeiting Methods

Some security holograms use a sculptured object. It is possible to create an imitation of the original artwork and a skilled holographer with a good lab is able to make holograms and copies that will pass careful inspection. Many security holograms are the so-called kinegrams. They are produced by a sequence of two-dimensional objects in different planes, which may be computer generated (Fig. 17.1). It is more difficult but also possible to reoriginate a kinegram.

Another method to counterfeit a hologram consists in contact copying a hologram as a one-step process. A holographic film is placed against the security hologram and illuminated with a laser. The film is developed and metalized. This method works for embossed hologram and for reflection holograms of the Denisyuk type. The result is a passable imitation of the original hologram. For mass production, a photoresist film can be used for counterfeiting embossed holograms. It is exposed, developed, and electroplated. The electroform is then used to emboss the counterfeited hologram.

Two-step copying is made by illuminating the original hologram with a laser and recording the diffracted light in a second hologram not in contact but at some distance (Fig. 17.6a). Then the second hologram is illuminated by a reconstruction laser beam. A replica of the image of the original hologram is produced (Fig. 17.6b). This image is used to make a duplicate of the original hologram. For an embossed hologram, the duplicate is recorded in a pho-

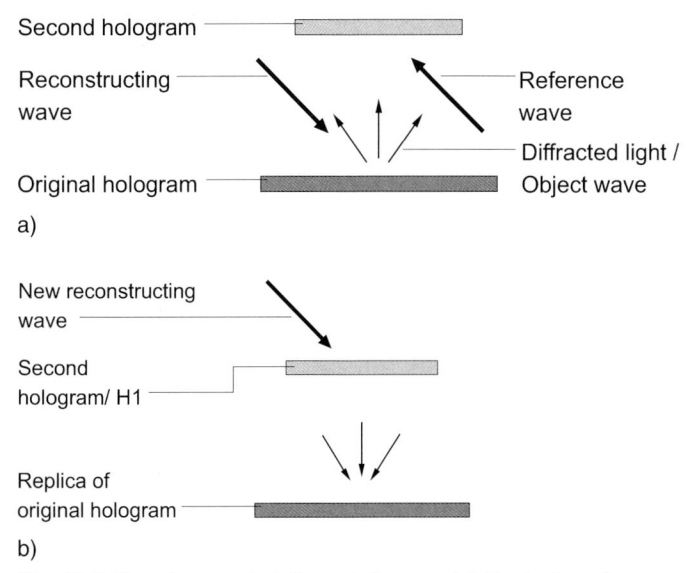

Fig. 17.6 Two-step counterfeiting a hologram. (a) Production of a second hologram (H1). (b) Production of a replica (H2).

toresist, which will be developed and electroplated. Using the electroform, counterfeit copies of the original hologram can be embossed. This process is more complicated than contact copying, but it produces better copies.

Mechanical copying is possible by uncovering the embossed surface and using it for the production of an electroform. Chemical processes may be used to dissolve different plastic layers of the embossed hologram.

17.2.2
Countermeasures

It is very difficult to develop a hologram which cannot be counterfeited. But companies for security holograms can make counterfeiting as difficult and expensive as possible so that they are practically secure. Several methods were proposed to complicate counterfeiting. First, the hologram may be produced by the stereogram process. In addition, hidden information may be recorded on the hologram, which cannot be seen by man but only by reading with a decoding device. Further variable information like serial numbers may be introduced in the hologram or the hot stamping process. Instead of one hologram several smaller holograms may be applied. Counterfeiting is also more difficult if special materials are used. The combination of several of these countermeasures results in secure holograms.

Problems

Problem 17.1 Estimate the depth Δx of the surface relief of an embossed hologram.

Problem 17.2 Is the master hologram in photoresist in Fig. 17.1a transmission or reflection hologram? Explain why this type of hologram is used as a master.

Problem 17.3 By hot stamping a surface phase hologram is produced. Explain why this type of structure is a transmission hologram. How can this hologram be used in reflection?

Problem 17.4 How is it possible to counterfeit (a) an embossed hologram and (b) a volume reflection hologram?

Problem 17.5 Why is it important to approximate a sinus grating of a computer-generated hologram (CGH) according to Fig. 16.7 using several production steps?

18
Holography and Information Technology

The optical processing and storage of information is considerably enriched by holography. Interesting fields of application are methods of analog pattern recognition (Sections 18.1 and 18.2) and holographic storage (Sections 18.2 and 18.3).

18.1
Pattern Recognition

18.1.1
Associative Storage

The basic equation (2.13b) of holography for the intensity distribution I within a hologram

$$I = |\mathbf{r}|^2 + |\mathbf{o}|^2 + \mathbf{ro}^* + \mathbf{r}^*\mathbf{o}$$

remains unchanged when exchanging the object wave $\mathbf{o}$ and the reference wave $\mathbf{r}$. When illuminating the hologram with the object wave $\mathbf{o}$ the reference wave $\mathbf{r}$ is reproduced, just like $\mathbf{o}$ is reproduced by $\mathbf{r}$. Thus one wave or information reproduces the other: this is the property of associative storage.

18.1.2
Pattern Recognition

The principle of holographic pattern recognition can be understood from the properties of associative storage. According to Fig. 17.1 a Fourier hologram of a transparent plane object is fabricated and is placed at its original location after development. If now the object is illuminated the object wave impinges on the hologram. Thus the reference wave is reconstructed and imaged spot like into the focal plane by lens 2. If the information of a wave matches the hologram the created signal is a diffraction limited spot at the output plane. Moving the object within the focal plane moves the spot accordingly in the focal plane of lens 2.

Holography: A Practical Approach. Gerhard K. Ackermann and Jürgen Eichler
Copyright © 2007 WILEY-VCH Verlag GmbH & Co. KGaA, Weinheim
ISBN: 978-3-527-40663-0

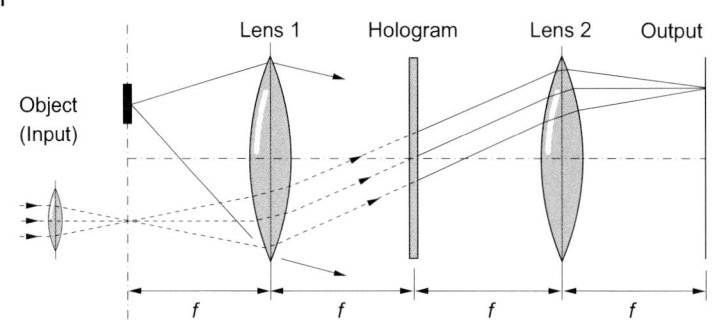

Fig. 18.1 Holographic setup for pattern recognition.

The setup in Fig. 18.1 can be used for automatic pattern recognition. The hologram with the sought-after information, e.g., a finger print or character, is placed in the common focal plane of the lenses. A plane object is placed in the focal plane of lens 1 and is illuminated so that the object wave with the sought-after information falls onto the hologram. Brightness and location of the light spots in the output plane deliver information about the degree of similarity and the position of the information on the object. The principle of pattern recognition is also called "spatial frequency filtering" since the frequency spectrum of the object wave is compared to the hologram.

18.1.3
Image Processing

Aberrations occur when using optical systems for imaging, which can be compensated by the described method of pattern recognition. A light spot is imaged with the setup and the resulting image is photographed. A Fourier hologram is fabricated from this photo which is placed into the setup according to Fig. 18.1. The images containing aberrations are used as the object and are illuminated with a laser. A compensated image is formed in the output plane. The operation can also be understood as follows: each (faulty) image point compared to its (also faulty) hologram. Consistency then leads to a point in the output plane that can be assumed to be faultless. Other methods of analog information processing like coding, multiplexing, and transformations are only of special relevance at the time [42].

18.2
Neuro Computer

In addition to digital computers also analogous systems that use holographic methods, e.g., neuro computers, are investigated [43]. These systems do not perform calculations but rather perform information comparisons and recognitions.

18.2.1
Recognition of Information

The principles of pattern recognition can be extended to three-dimensional storage media. An example is shown in Fig. 18.2: In a light sensitive crystal multiple holograms are recorded on top of each other with slightly angled reference waves. This system serves the purpose of comparing the content of the memory with unknown or fragmentary information which is impinging as a light wave onto the hologram. If the structure of the light wave matches the recorded hologram the according reference wave is reconstructed. This reference wave falls onto a phase conjugated mirror which reflects the impinging wave into itself. That way the according hologram is read out. The process runs analog if the entered information only partially matches the recorded hologram. Missing or erroneous information is completed and corrected.

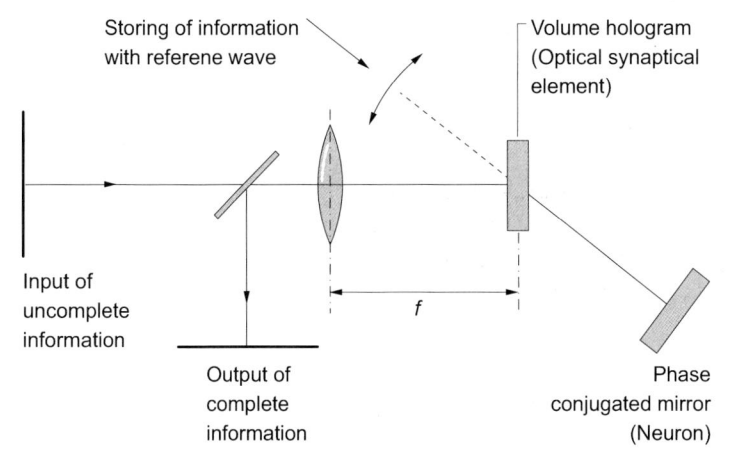

Fig. 18.2 Holographic system for reading and correction of information, optical associative memory, optical neural network.

Phase conjugated mirrors are based on principles of nonlinear optics; they only reflect above a certain intensity threshold. If the similarity between the illuminating and the recorded information is too low the according weakly reconstructed reference wave is not reflected by the mirror. The phase conjugated image acts as a neuron, which represents a logical element with multiple analog inputs but with only one output. The function of the hologram can be thought of as a "synaptical element" which links information in parallel.

18.2.2
Phase Conjugated Mirrors

A phase conjugated mirror has two important properties: An impinging wave is mirrored 180° independent of the angle of incidence and the phase plane of the reflected wave acts like the conjugated image in holograms. Such mirrors are fabricated as realtime holograms by illuminating a photorefractive or similar material with coherent light. The impinging wave represents the object wave, the hologram is created by interference with a plane reference wave. At the same time another plane reference wave is used which travels in the opposite direction of the reference wave and reads out the hologram. A pseudoscopic image is created which is formed by the mirrored phase conjugated wave. Beneath its application in neuro computers the phase conjugated mirror is a perfect laser mirror since it does not require adjustments and compensates for inhomogeneities within the laser medium.

18.3
Digital Holographic Memories

The importance of optical memories lies in their high storage density, fast access time, and contact-less reading possibility. Holographic memories have additional advantages: The information of a single bit is spread over a wider area so that small defects in the storage medium do not delete single bits but only lower the signal-to-noise-ratio and the information can be read out parallelized. With volume holograms the information can be stored in three-dimensional media with densities up to 1 bit per cube of the recording wavelength [44, 45].

18.3.1
Stack Organized Memories

In the following the principles of a stack organized holographic memory will be described. The information of a data page containing ca. 1 Mbit is represented on a data mask. The mask is similar to a matrix-like spatial light modulator which represents a two-dimensional transparent object (Fig. 18.3). It can be made of a liquid crystal layer or an electro-optic material.

The data mask is illuminated by a laser (e.g., Nd–YAG, frequency doubled). It creates an object wave that is focused onto a storage medium with a diameter of ca. 1 mm using a lens. This lens is omitted in Fig. 18.3. By switching the direction of the illuminating laser beam (reference wave) it is possible to select e.g., 10,000 positions on the storage medium. Each position represents one data stack.

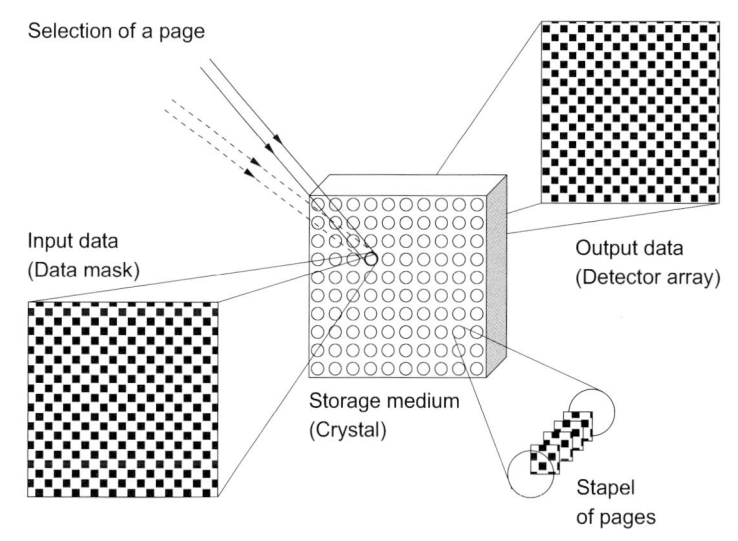

Fig. 18.3 Principle of a stack organized holographic memory. The lenses between data mask and storage medium and between storage medium and detector array are omitted.

The storage medium is, e.g., made of a photorefractive crystal of the size $10 \times 10 \times 0.5$ cm^3 whose refractive index depends on the exposure (Section 14.6). A volume-phase hologram is created within this material. More than 100 holograms can be recorded on top of each other on each position of this storage medium (data stack). Before each recording the reference beam is tilted by approximately 0.25°.

For reading out the information from the holographic memory the selected stack is illuminated with the reconstruction wave that resembles the reference wave. By tilting the angle of incidence the desired page is selected. Here the Bragg condition ensures that each page can be read separately. An image of that page is created that is imaged onto a detector array (Fig. 18.3). The information can be processes serially or parallel. The deletion of the information is done by homogeneous illumination or by heating. Problems prevail with the nondeletable readout.

18.3.2
Stack Organized DVD

The principle of the stack organized holographic memory according to Fig. 18.3 can also be applied to a rotating storage medium. A photopolymer disc of about 1 mm thickness can be used as a holographic DVD. The selection of the stack on the disc can be made by radial movement of the disc or the optical system. The page in a stack is selected by the angle of incidence of the reference beam which is equivalent to the reading beam.

18.3.3
Microholographic DVD

In Fig. 18.4 a DVD is shown where 1 bit is represented by a holographic micrograting. The phase grating is produced by blue diode laser radiation which is focused perpendicular to the surface into the interior of a thick photorefractive holographic layer. The transmitted laser beam is reflected by a spherical mirror (Fig. 18.4a). The transmitted and reflected beams interfere and an interference grating is created with a grating period of half the wavelength in the medium. For a wavelength of 400 nm and an index of refraction of 1.5 a grating period of 133 nm arises. Since the laser beam is focused the diameter of the grating is of the order of the wavelength of the laser depending on the focal length and the lens diameter. Thus a typical spot size of about 400 nm occurs. The micrograting has a significant index of refraction modulation only within the Rayleigh length of the focused laser beam of about 10 µm. Each bit is written on the rotating disc in about several ns. For reading the holographic bits an arrangement according to Fig. 18.4b is used. The mirror is removed and the writing laser beam is reflected on the micrograting.

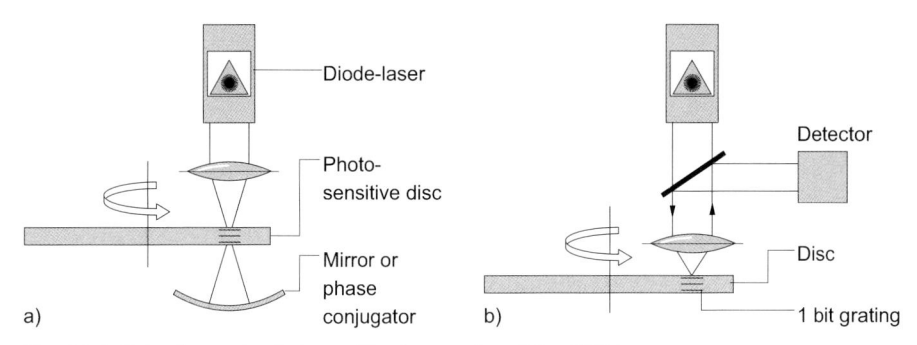

Fig. 18.4 Setup for a microholographic storage disc of the DVD type: (a) writing bits as a micrograting and (b) reading by Bragg reflection from the microgratings.

The microholographic disc has the advantage that multiplexing with different methods is possible. First, different wavelengths may be used writing several bits in the same volume. A typical value at the moment is about three wavelengths. Second, in a thick holographic layer the grating can be written independently in different planes, e.g., 16 planes. Thus, the storage density may be nearly 3×16 times higher than in a normal DVD. Microholographic storage requires simpler hardware than other types of holographic storage and may be compatible with DVD technology.

Problems

Problem 18.1 How is it possible that a holographic neuro computer recognized, completes, and corrects information?

Problem 18.2 Describe a holographic phase conjugated mirror using a figure.

Problem 18.3 What is the upper limit for the storage density of a holographic volume memory with an index of refraction of $n = 1.5$ with a volume of $V = 1$ cm^3 using a laser with $\lambda = 400$ nm.

Problem 18.4 Calculate the focal spot radius w_0' and Rayleigh length z_R of a 400 nm laser beam (radius $w_0 = 1.5$ mm), which is focused by a microscope objective (focal length $f = 4$ mm) within a microholographic DVD (index of refraction $n = 1.5$). What is the grating period?

Problem 18.5 Estimate the distance Δz of planes with different holographic gratings in a holographic DVD using the data of Problem 18.4.

19
Holography and Communication

19.1
Holographic Diffuser Display Screen

19.1.1
Lambertian Diffuser

Conventional diffusers such as ordinary frosted glass, ground glass, milky glass, opaque plastics, etched plastic, or metals scatter incident light in all directions. They achieve their diffusion by means of surface roughness which is difficult to control. In many cases the scattered intensity or power density $I(\theta)$ in diffuse reflection or transmission through a diffuser depends on the angle θ and is given by the equation of Lambert

$$I(\theta) = I(0)\cos\theta, \tag{19.1}$$

where $I(0)$ is the intensity at $0°$ (Fig. 19.1). The scattered intensity is high in the forward and backward directions. It decreases considerably with the observation angle θ which is unfavorable for many applications, e.g., for a diffuser screen for projection with a beamer.

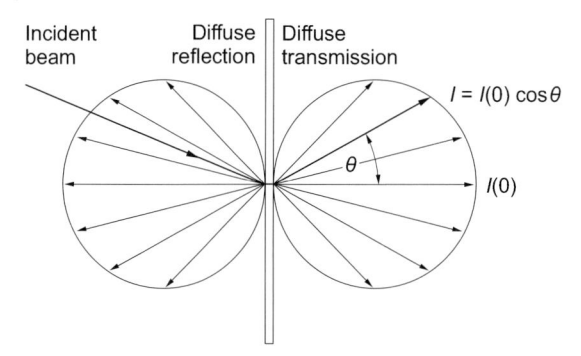

Fig. 19.1 Diffuse reflection and transmission for a Lambertian diffuser.

Holography: A Practical Approach. Gerhard K. Ackermann and Jürgen Eichler
Copyright © 2007 WILEY-VCH Verlag GmbH & Co. KGaA, Weinheim
ISBN: 978-3-527-40663-0

Using holographic methods it is possible to built diffuser screens with an arbitrary angular distribution of the scattered intensity $I(\theta)$. In the following only diffusers in transmission are discussed because at the moment they have more applications than diffusers in reflection.

19.1.2
Holographic Diffuser Screen [46, 47]

A normal transmission hologram produced by two plane or spherical waves acts as some kind of diffuser. If such a hologram is illuminated the light is diffracted into a special direction depending on the construction of the hologram. If the white light is used a strong chromatic aberration occurs which prevents technical applications as a diffuser screen.

Holographic diffuser screens with negligible chromatic aberration can be constructed according to Fig. 19.2a. A normal diffuser is illuminated by a laser beam and the scattered radiation represents the object wave. The reference wave from the same laser enters the holographic plate from the same side as the object wave. Thus a transmission hologram of the diffuser is obtained that acts as a holographic diffuser screen.

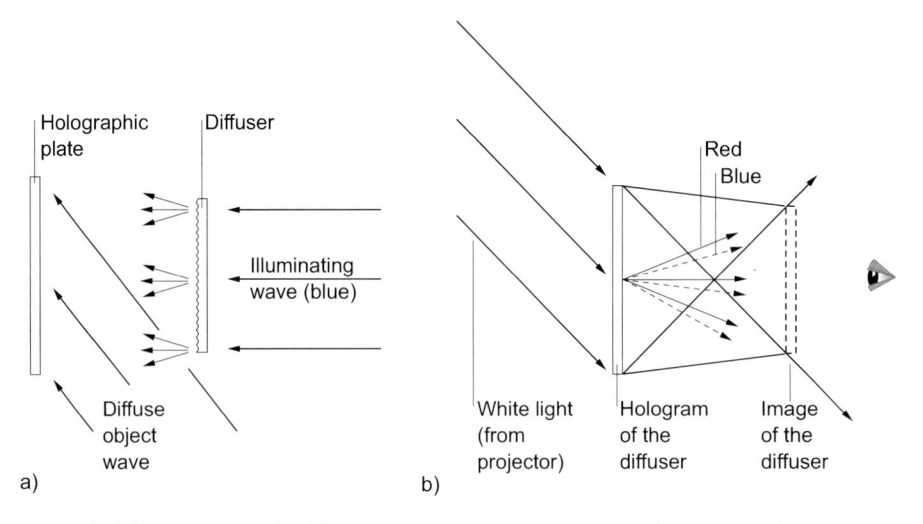

Fig. 19.2 (a) The holographic diffuser screen is a transmission hologram created by an object wave of a laser illuminated normal diffuser screen and the reference wave. (b) A projector (beamer) focuses an image on the holographic diffuser screen. The observers eyes are focused on this image. Chromatic aberration is omitted and the image is seen in its original colors. (The hologram creates images of the diffuser which are shifted for each wavelength. However, close to the optical axis all colors are mixed.)

For practical application the holographic screen is illuminated with white or colored light from a beamer or another projector according to Fig. 19.2b. This light represents the conjugated reference wave with an opposite direction of propagation with respect to the reference wave. The projector focuses the projected image on the holographic screen. The holographic screen acts as a transmission hologram diffracting the light in the direction of the observer. An image of the normal diffuser is reconstructed with chromatic aberration. Due to diffraction the positions of the images for each color are shifted. However, Fig. 19.2b shows that different colors are mixed to white light in the largest part of the image of the diffuser. Color smearing occurs only at the edges of the image. In the center all colors coming from a point of the hologram are mixed. The eye is focused on the projected image on the holographic screen and the colors of the projected image are seen perfectly.

For producing holographic diffuser screens with large areas the production can be made step-by-step according to Fig. 19.3a. A small normal diffuser screen and a reference beam with small diameter are used and the holographic plate is exposed successively in small areas. For this purpose the diffuser and the reference beam are moved over the holographic plate. Figure 19.3b shows that one small hologram produces the image of the small diffuser. The other small holograms are not shown but they act in a similar manner.

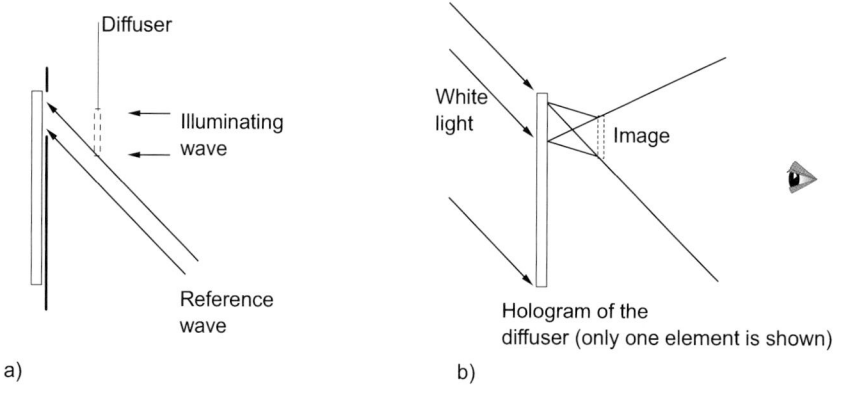

Fig. 19.3 Production of a large holographic diffuser screen: (a) the holographic plate is exposed step-by-step superimposing the radiation from the small diffuser and the reference wave and (b) the complete holographic diffuser plate is illuminated by the light of a projector. Each small hologram on the holographic diffuser produces an image of the small diffuser.

For the production of holographic diffuser screens dichromated gelatin or silver halide materials can be used. For mass production the methods of embossed holography can be applied.

19.2
Holographic Display [47]

The holographic diffuser screen diffracts the light from a projector into a well-defined direction and solid angle. Thus diffuser screens for stereoscopic 3D projection can be designed modifying the methods of Section 19.1.

For stereoscopic viewing different images have to be seen by the right and left eye. Usually two images are projected on a screen and the two images are separated by special glasses. The most common method uses two images with different polarization filters in the glasses. Another method uses two images with red and green colors and the corresponding filter glasses. It is also possible to project the two images successively and separate them with a gated shutter in front of each eye.

Figure 19.4 shows a special holographic 3D screen which can be used without glasses. Two images for the right and the left eye are projected on this screen under different angles by the projectors 1 and 2. The 3D screen diffracts the light from two projected images into different directions. The observer with his eyes in a position shown in Fig. 19.4 focuses his eyes on the holographic 3D screen and sees a different image with both eyes. Thus autostereoscopic 3D vision is achieved.

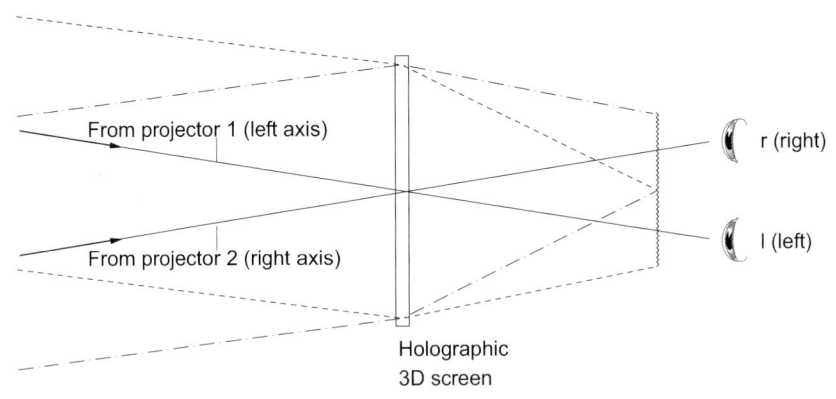

Fig. 19.4 Holographic 3D screen for autostereoscopic viewing of two projected images, one for each eye.

The holographic 3D screen consists of two multiplexed holograms. The object waves for these holograms are created by two normal diffusing screens which are positioned side by side. Two reference waves with different angles of incidence are used, one for each normal diffuser screen. The principle of the production is similar to Fig. 19.2a using two diffusers and two reference beams. The system of Fig. 19.4 consists of one projected view and only one observer is admitted with his eyes in the exact position. By multiplexing more than two holograms on the 3D screen multiple projected viewing regions can be achieved. A 3D screen with eight projected views can be obtained, four for the left and four for the right eye. The action of such a 3D screen is shown in Fig. 19.5.

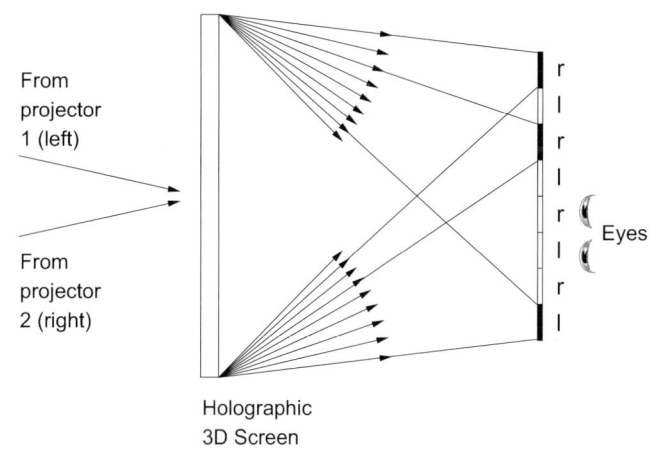

Fig. 19.5 Holographic 3D screen with eight views of two projected images. (r = right eye, l = left eye)

19.3 Holographic TV and Movies

It is not possible to produce holograms from scenes which are illuminated by daylight because it is incoherent. Thus holographic TV and movies must include stereoscopic methods. The scene is shot in the known manner from different perspectives simultaneously using 2D film material. From this almost conventional stereo movies a series of holograms is created. The holograms produced from the stereo movie can be projected directly on the photosensitive layer of an electronic camera (without objective). An electronic camera is able to convert optical information into electrical impulses. However, the resolution of nowadays CCD or CMOS cameras has to be improved by about one order of magnitude. The electronic image signals are stored using TV technology and are transmitted later to be displayed on a special TV set (Fig. 19.6).

The amount of data that is stored in a hologram can be calculated for a 10 cm × 12.5 cm hologram with a vertical pixel density of 1600 mm^{-1} and a horizontal density of 800 mm^{-1} to roughly 5×10^{10} pixels. The common frame rate for transmission is 30 frames per second. This flow of data cannot be handled by any technology available nowadays. The amount of data that can be handled in actual electronic image transmission is at least by a factor of 10^5 smaller. However, there are possibilities to reduce the amount of data

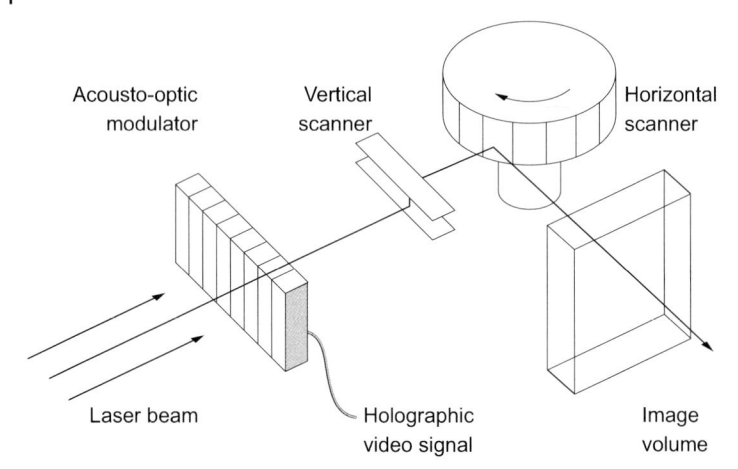

Fig. 19.6 Schematic setup of a holographic video system. One line is displayed as a moving phase hologram in an acousto-optical modulator and is scanned by a laser beam. Scanner and polygon mirror have the purpose of composing the image. Lenses in front of and behind the polygon mirror are omitted for clarity.

in holographic TV. It is known from rainbow holography that the vertical parallax can be omitted without a loss in three-dimensional impression for the observer. Additionally the information density in a vertical line is matched to the resolution of the human eye.

19.3.1
Holographic TV

Benton has proposed a holographic video system for the display of images of holograms (Fig. 19.6) [48]. Using the methods described above the amount of data for a 24×36 mm^2 size hologram with an angle of view of $12°$ is reduced to 192 lines with 32,000 pixels each. To reduce the size of the needed screen the display is performed sequentially similar to normal TV. The hologram is displayed line by line. For this an acousto-optic modulator is used into which the hologram of one line is written as a phase grating by a piezoelectric element. The hologram line propagates with the speed of sound through the longish laser beam. This way the holographic image of a line is created.

From the moving images of the holographic lines a three-dimensional image is created using a vertical scanner and a rotating mirror. The speed of the mirror is matched to the speed of sound in the acousto-optical modulator. In this experiment it was shown that holograms can be displayed by means of electronic media although by dropping part of the information and limiting to very small objects.

19.3.2
Alternative Methods

Experiments were also done to record holograms with the use of electronic cameras and to display them with liquid crystal displays (LCD). These are illuminated with a laser for reconstruction. By now the results are not satisfactory due to small resolution of available components. If the development leads to components that exhibit a resolution of 1000 to 3000 lines per millimeter there could also be a future for holographic TV.

Another experimental holographic system uses digital micromirror devices (DMD) as a holographic layer. These DMDs are usually used in video projectors. Each quadratic micromirror has a typical length of several µm and thus it is possible to electronically write a hologram within a DMD. Such a DMD device can be illuminated with a laser beam and a holographic 3D image can be created. This image is projected on a volumetric screen. This screen consists of parallel layers of thin LCD panels which can be switched clear or opaque by an electronic signal. At a given time only one panel is opaque and the other panels flash on and off in quick succession. Thus by light scattering the whole holographic 3D image can be seen [49].

Similar displays exist today but they work without holography. Each LCD panel shows a slice of a 3D image in a quick sequence resulting in a 3D image. The picture is almost the same as the one given by the holographic DMD system. However, it can be shown that the holographic system requires only the bandwidth of nowadays TV signals whereas other systems need the normal bandwidth multiplied by the number of LCD panels.

19.3.3
Holographic Movie

The conventional procedures for the production of three-dimensional movies use anaglyphic methods or work with polarized light. Both methods use spectacles with different colored glasses or polarizing filters and are derivatives of older stereoscopic techniques: Each eye is presented a different perspective of the shot scene. From these two-dimensional information the human brain reconstructs a three-dimensional image. A stereoscopic method without spectacles was described in Section 19.2.

For more than 20 years now it is attempted to also produce three-dimensional holographic movies. The advantage is the real three-dimensional display without the use of any other viewing aids. From the theoretical point of view such a production can be done without any problems. The movie passing the projecting optic consists of a series of Fourier holograms. For these the image position does not change when the holograms are moved (Section 4.5). However, two basic problems hamper the presentation of a holographic movie. Firstly, the display of holograms is limited to a small viewing angle. This opposes the demand to present the movie feature to a larger audience. Secondly, it is not possible to project the holographic movie onto a screen without the loss of the three-dimensional vision.

19.3.4
State-of-the-Art

In the last decades only small technical progress was made in the development of holographic movies. Jacobson et al. [50] did produce a movie of about one minute length but only one observer at a time was able to view it. Therefore, Komar [51] focused his works on the problem of the projection of holograms. He designed a holographic screen which acts as an HOE like a superposition of elliptical mirrors with different eccentricities. With this HOE as a projection plane the real image of the holographic movie which is placed in the common focal point is imaged into the other spatially separated focal points. However, also with this method the number of viewers stays small. A satisfying solution to this problem does not exist yet. Therefore, the known 3D methods in movie technology will not be replaced by holographic ones in the near future.

Also untreated is the question of holographic color movies. Although no principal problems exist all known true color methods are very expensive.

As long as the objects and scenes to be shown in the movie are on the laboratory scale the techniques described in this book are applicable. It gets difficult when a large area like a landscape or a street scene is to be recorded. An illumination with laser light is obviously not possible. This problem also plays a role for holographic TV and was already described in this section.

Problems

Problem 19.1 Estimate the angular range $\pm\theta$ of a holographic diffuser screen, which is produced according to Fig. 19.2?

Problem 19.2 What is the data rate for a holographic TV screen with an area of 1.2 m $\times$ 1.0 m? The horizontal bit size is 1 µm and the vertical size is 5 µm.

Problem 19.3 Is it possible to produce a hologram for a large scene in nature?

Problem 19.4 Describe a holographic 3D screen for the display of stereo images.

Problem 19.5 What are the principles of a normal 3D movie?

Problem 19.6 A holographic diffuser consists of a grating structure diffracting the light from every pixel into a limited range of angles (Fig. 19.2). How is it possible that chromatic aberration does not disturb the image seen on the screen?

20
Holography – Novel Art Medium

20.1
Artistic Holographic Works

Form and color are two essential means of expression in arts. Therefore, it is not surprising that holography early caught the interest of artists. With the standardization of the holographic procedures that are presented in this book the necessary technical "know how" is governable for everyone interested. Colors and presentation of forms using holography got accessible for a group of people who were more interested in the artistic composition rather than in the scientific basics of holography. Besides that many holography laboratories exist that offer the technique as a service.

Design and coloring have reached new heights by the use of holography. It is possible to create three dimensional images without having to refrain from effects that are known from painting like the distribution of light and shadows. Additionally, new possibilities arise and more variety in the choice of perspectives. Also a new use of color was added. An image can be designed in different color distributions simultaneously. Different techniques on how colors can be realized in an holographic image were presented in the preceding chapters.

20.1.1
Regarding the Critics on the Medium Holography

Holography was always criticized for being more a technical medium which can only be used for technical applications and that its artistic value is secondary – if present at all. In fact the complex physical and chemical working methods have to be pointed out whose knowledge is inevitable to realize colors and shapes in a desired way in a holographic artwork. However, technical knowledge is needed in a lot of artistic media such as, e.g., copper plate engraving, painting, or sculpture to name just a few.

Marc Piemontese in this context refers correctly to Charles Baudelaire who in his time said that photography "brings industry to arts" [52]. In fact, arts

Holography: A Practical Approach. Gerhard K. Ackermann and Jürgen Eichler
Copyright © 2007 WILEY-VCH Verlag GmbH & Co. KGaA, Weinheim
ISBN: 978-3-527-40663-0

and technology never were opposites; especially photography developed into an accepted artistic medium. Margaret Benyon, one of the most well-known holographic artists in England noticed a growing restraint of the classical art scene from holography. So holography is counted as one of the "new techniques" in the same line as video, movies, TV, and computers. This makes clear, as she stresses out, how dissociate the artistic establishment is from holography. For sure holography belongs to these novel techniques and therefore opens up new dimensions for other media. One example for that is the installation "H.O.E.-TV" or "Diffracted Wall" by V. Orazem and T. Lück. In H.O.E.-TV under the title "Radical Holography" the attempt is made to reveal the artistic content of holography. The holograms are reduced to their basic optical elements (HOE). A TV screen is used as a light source displaying white noise or computer animations [53].

20.1.2
Examples of Art and Holography

The question if holography is an artistic medium was long unanswered. The impact in the area of arts can be shown best in some examples. A definite style cannot yet be found in this young section of the fine arts. Perhaps the variety is an expression of the new possibilities. The exhibitions by D. Jung (Kunsthochschule für Medien, Cologne) show one way. Abstract shapes and a virtuous play with colors characterize his artworks. He consequently uses the possibilities that holography offers, e.g., the interplay of colors which cover highly variable all parts of the spectrum in different areas of the image during the motion of the observer.

Also R. Berkhout, P. Newman, and V. Horazem do not portrait real objects in their holograms but deal with the light phenomena. In the holograms of Berkhout space and light landscapes appear in which the observer can take different perspectives. Newman guides the laser beam through lenses, ground and broken glasses, scattering objects and other materials during the recording. Thus interesting and unusual light effects are created.

Some artists such as Ana M. Nicholson, H. Casdin-Silver, M. Crenshaw, S. Dinsmore, M. Benyon, J. Kaufman, or S. Cowles are dedicated to more concrete presentations in holography. D. Randazzo combines computer, video, movie, and photography with holography. Ana M. Nicholson mainly works in the portrait area. A complete list of all holographic artists and styles is not possible within the scope of this book.

Margaret Benyon [54] worked a lot in the area of portrait holography. There she also adapts other graphical techniques and joins them with holography. By this mixture of two-dimensional and three-dimensional elements a special tension is created in her artworks. Other more commercial works on portrait holography were produced by R. and B. Olson.

A large circle of artists works with installations. Here the border between arts and kitsch is very often crossed. From the artistic point of view it is not very convincing to use holography only as a three dimensional imaging method. Photography also is not by itself an artwork. In the artistic field of installations the discussions are controversial and the opinions are very different. Especially important contributions are being made by Doris Villa and Dan Schweizer from New York and by Alexander from California. Every now and then attempts are made on creating very large installations up to the scale of architectural designs. The "Quelle mit sterbenden Blättern" is an installation of 3 m × 3 m that was presented at the Technical Fair in Flanders in 1991. One large installation project was presented by Paula Dawson (Australia) at the "International Symposium on Display Holography" in Lake Forest near Chicago [55]. The installation "You are here" is to be mounted at a shore line in Australia and is supposed to be reconstructed at moon light. The observer of the next million years according to Dawson is then able to recognize the short and long term changes in the landscape.

Also a topic of growing interest is "Architecture and Holography" (Section 21.3). Examples for this are "Transponder," a work by D.E. Tyler at the scientific center of the University of Nebraska, and a large hologram "Eye Fire" by M. Bleyenberg in Bonn (German Research Association DFG). The examples that are completed in the German region by the works of V. Orazem and D. Öhlmann, Hochschule für bildende Kunst in Braunschweig, and by R. and B. Rosowski – the latter due to their controversially discussed image montages – are supposed to give an insight although not a complete overview to the variety of artistic works in the field of holography.

The fine arts will have to deal more serious with the area of holography then it was done until now. Both sides would benefit from this. The connection of technique and art in holography is done and developed further by the symposia which are held for more than 20 years now at the Lake Forest College. Other important symposia take place at the annual conference "Photonics West" in San Jose. These events bring together artists from different classical art fields with the holography is still to come.

Due to the direct and successful application in advertising the cooperation of graphics and holography is more natural than in other areas of the fine arts. Graphic works are directly applied in the industry and even reach the design of objects of daily life.

20.2
Portrait Holography

In principle, two systems for portrait holography are used. The *first* is a normal hologram with a nanosecond pulsed laser. Usually first a master transmission hologram (H1) is made and then in a second step a reflection copy (H2) [56]. However, it is also possible to produce a reflection portrait hologram in one step [57]. In this case the hologram is seen behind the holographic plate. The *second* system is a three-step multiplexing process starting from a series of photographs from different views of the person. In the following first the system with a master (H1) and a reflection copy (H2) is described and then the multiplexing method.

20.2.1
Lasers for Pulsed Portrait Holography

In the beginning of portrait holography only Q-switched ruby lasers with a pulse energy of about 1 J and more were available. The laser is equipped with an oscillator, an amplifier, a Q-switch, and an etalon yielding a coherence length of about 1 m. The pulse width of $t = 10$ ns is so short that the movement of a person with a velocity of $v = 1$ cm/s results in a displacement of $s = v \cdot t = 10^{-4}$ µm. This is much less than the distance of the interference fringes in a hologram which are in the µm region and movements are not a problem in portrait holography. The wavelength of the ruby laser is 694 nm. The red laser radiation penetrates the skin by several mm before being backscattered. This results in unnatural images. Thus it is advantageous to use some special makeup, e.g., cosmetic face powder and green lipstick.

Later frequency doubled Nd:YAG lasers became available. Typical technical specifications are similar to that of the ruby laser. However, the wavelength is 532 nm resulting in a green radiation. The penetration into the skin is only several 0.1 mm and the images are more natural.

20.2.2
Master (H1) and Reflection Copy (H2)

Because of the short exposure time mechanical stability and movement problems are eliminated. Figure 20.1 shows the arrangement for producing a master transmission hologram (H1) of a person. It is important to avoid eye injuries from the laser radiation. Thus the person is not irradiated by the direct laser radiation but by the radiation scattered from a glass plate diffuser. In the setup of Fig. 20.1a two diffusers are used for a more homogeneous illumination. While the direct laser radiation is focused on the retina on a very small area with a diameter of about 10 µm the diffuser is imaged on a larger area with a diameter D' which is a function of the diameter of the diffuser D, the distance of the diffuser x and the focal length of the eye $f = 2$ cm. With $D = 20$ cm and $x = 100$ cm the result is $D' = Df/x = 0.4$ cm. Thus the laser energy is distributed over a large area reducing the risk considerably. A calculation of the maximum permissible exposure (MPE) and the diffuser radius is given in the following subsection. Typically diffusers with a diameter of about 20 cm and more are used and the scattered light is not dangerous for the opened eye for pulse energies of up to 1 J.

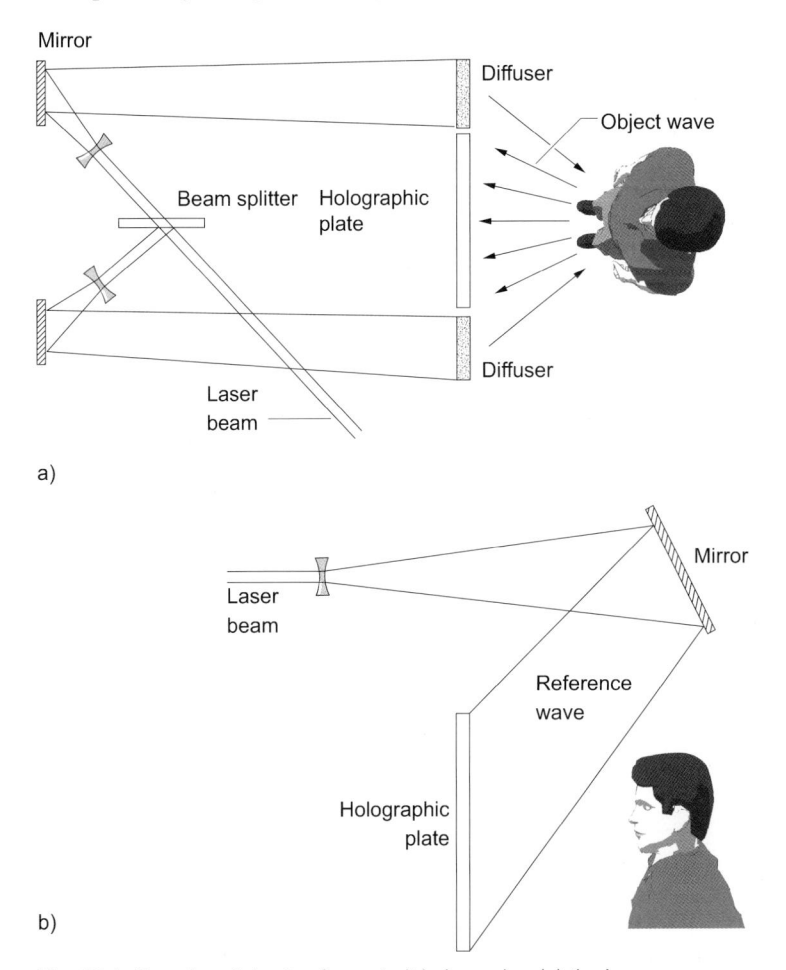

Fig. 20.1 Experimental setup for portrait holography: (a) the laser beam is directed to two diffusers which illuminate a person. The radiation scattered from the person is the object wave and (b) the reference wave is directed from above to the holographic plate.

The object wave is the radiation scattered from the person which is illuminated by the laser radiation from the diffusers (Fig. 20.1a). The reference wave is directed from above on the holographic plate (Fig. 20.1b). The reference wave is most dangerous to the eyes. Even the reflection from the holographic plate will produce an eye injury if it is directed into the eye. Thus the person must be carefully protected from the reference wave. A good film material for production of a transmission master hologram (H1) is produced by the Slawish company.

From the H1 hologram a reflection copy (H2) can be produced as described in Chapter 11. The most suitable laser for copying seems to be the cw frequency doubled Nd:YAG laser with a power of several 100 mW and a coherence length of many meters. For reflection holograms a different film material may be used, e.g., Millimask from Agfa which is the former Agfa 8E56.

20.2.3
Eye Safety Calculations

For portrait holography pulsed ruby lasers or even better frequency doubled Nd:YAG lasers are used. Typical pulse widths are in the 10 ns region with pulse energies of about 1 J. These lasers are extremely dangerous for the eye and safety considerations are necessary based on the MPE (maximum permissible exposure) published by the American National Standards Institute ANSI [58]. The maximum energy density H_{MPE} measured at the entrance of the eye close to the cornea for a wavelength between 400 nm to 700 nm and a pulse width between 10^{-9} s and 1.8×10^{-5} s is

$$H_{\text{MPE}} = 5 \times 10^{-3} \, \frac{\text{J}}{\text{m}^2} = 5 \, \frac{\text{mJ}}{\text{m}^2} \tag{20.1a}$$

for a direct laser beam. Equation (20.1a) gives the threshold for damaging the retina looking directly into the laser beam. This energy density is too low for portrait holography. Introducing a diffuser higher energy densities are admitted because the diffuser is imaged on the retina resulting in a relatively large image reducing the power density. The MPE value for diffuse scattered laser radiation is given by [58]

$$H_{\text{MPE}} = 5 \times 10^{-3} \cdot C_6 \, \frac{\text{J}}{\text{m}^2} = 5 \cdot C_6 \, \frac{\text{mJ}}{\text{m}^2}. \tag{20.1b}$$

The function C_6 depends on the angular subtense α of the diffuser (Fig. 20.2a) and is given in the tables of ANSI [58].

Large diffuser: In the following a "large" diffuser is assumed with $\alpha \geq 100$ mrad. In this case we get $C_6 = 66.7$ and $H_{\text{MPE}} = 0.33 \, \text{J/m}^2$ measured at the cornea [58]. According to the rules of ANSI the measurements of the power density have to be performed using a detector and a circular aperture limiting the angle of acceptance to $\gamma = 100$ mrad independently of α (with $\alpha \geq 100$ mrad). The situation is demonstrated in Fig. 20.2a: α describes the

20.2 Portrait Holography

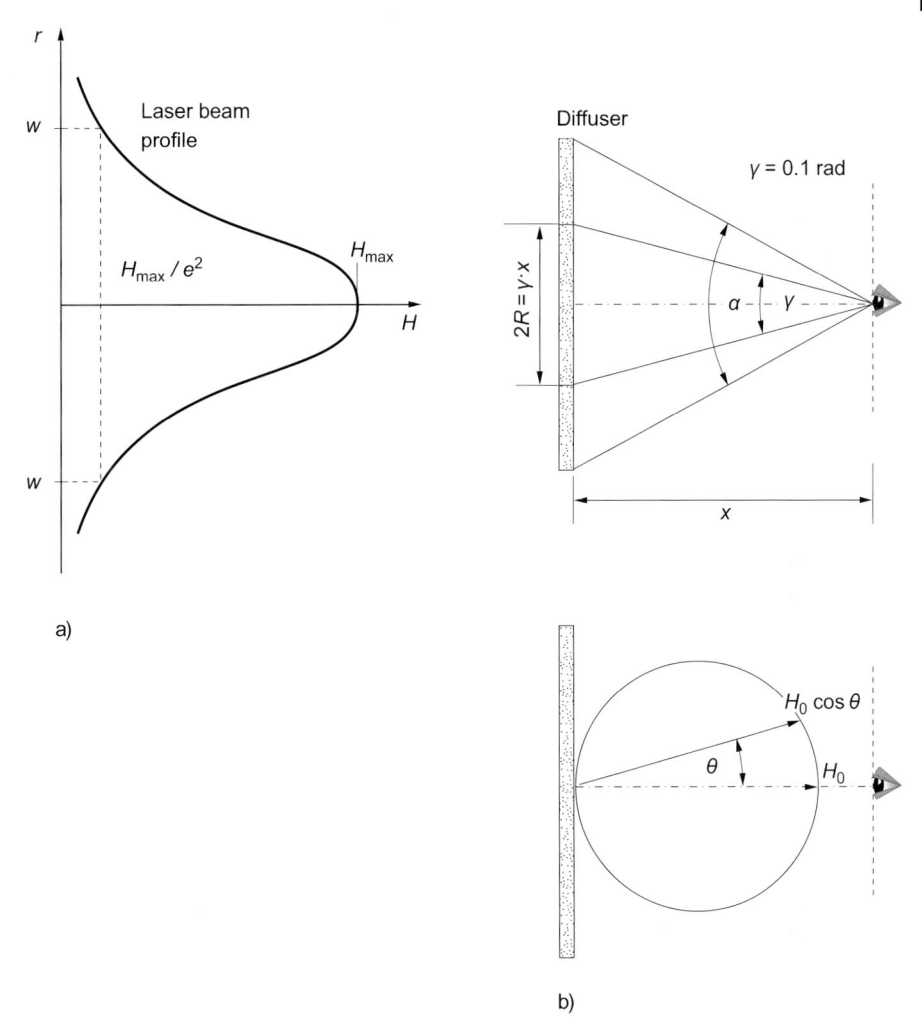

Fig. 20.2 Geometry for safety calculation according to ANSI: (a) the geometry of the diffuser screen and the energy density H of of the laser beam are shown and (b) characteristics of a Lambertian diffuser.

angular subtense of the diffuser and $\gamma = 100$ mrad the angle of acceptance for the pulse energy measurement for determining the energy density H_0. The laser safety rules by ANSI give $H_{\text{MPE}} = 0.33 \text{ J/m}^2$ and the mentioned description to measure the experimental value H_0. For eye save conditions we must have $H_0 \leq H_{\text{MPE}} = 0.33 \text{ J/m}^2$.

In the following estimations are discussed which give some more details about the experimental conditions for eye save portrait holography. We suppose that the diffuser is described by the law of Lambert, giving the energy density H as a function of the scattering angle θ, the pulse energy Q_{measured}

and the distance x from the diffuser according to Fig. 20.2b:

$$H = H_0 \cos\theta \quad \text{with} \quad H_0 = \frac{Q_{\text{measured}}}{x^2 \pi}. \tag{20.2}$$

The value $H_{\text{MPE}} = 0.33 \, \text{J/m}^2$ is only valid for a large screen ($\alpha \geq 100$ mrad). This means that the laser beam diameter is larger than the area for the energy measurement with a diameter of $2R \approx \gamma x$ (Fig. 20.2a). We assume that the energy density of the laser beam in the area with a diameter of $2R$ is approximately equal to the maximum value in the beam H_{max}. For a Gaussian beam H_{max} is given by the pulse energy Q and the beam radius w ($\frac{1}{e}$-value):

$$H_{\text{max}} = \frac{2Q}{w^2 \pi}. \tag{20.3}$$

In this equation the total pulse energy is given by Q. The energy Q_{measured} is the pulse energy measured according to the rules of ANSI within the area with a diameter of $2R$. For the estimation of Q_{measured} we assume a constant value of H_{max} over the area with the diameter of $2R$, yielding

$$Q_{\text{measured}} = H_{\text{max}} R^2 \pi = \frac{2QR^2 \pi}{w^2 \pi} = \frac{Q\gamma^2 x^2}{2w^2}. \tag{20.4}$$

Thus we obtain the energy density at the cornea H_0:

$$H_0 = \frac{Q_{\text{measured}}}{x^2 \pi} = \frac{Q\gamma^2}{2w^2 \pi} \leq H_{\text{MPE}} = 0.33 \frac{\text{J}}{\text{m}^2}. \tag{20.5}$$

From Eq. (20.5) we get for the laser beam radius for eye save operation

$$w \geq \sqrt{\frac{Q\gamma^2}{2\pi H_{\text{MPE}}}} = \sqrt{\frac{Q \times 10^{-2}}{2\pi 0.33} \frac{\text{m}^2}{\text{J}}} = 7 \, \text{cm} \quad \text{for} \quad Q = 1 \, \text{J}. \tag{20.6}$$

Thus for a pulse energy of $Q = 1$ J we get a laser beam radius of $w \geq 7$ cm. The radius of the diffuser is given by $R \geq \gamma x/2 = 0.05 \, x$, where x is the distance of the person from the diffuser. It is also necessary that the radius w of the laser beam is chosen larger than the radius of the area of measurement given by the ANSI rules: $w \geq \gamma x/2 = 0.05 \, x$. If these conditions cannot be complied with it is also possible to perform the estimation for smaller diffuser screens. In this case a new calculation of C_6 has to be made according to the ANSI tables [58].

> Note: It is extremely important not to rely on the calculations but it is necessary to measure the power density H_0 at the position of the eyes and compare the value with $H_{\text{MPE}} = 0.33 \, \text{J/m}^2$. In many cases diffusers do not have a Lambertian distribution but they show a stronger scattering in the forward direction. In these cases the laser beam radius w has to be larger than the calculation according to Eq. (20.6).

Reference beam: It is extremely important that the reference beam does not enter the eyes during portrait holography. The MPE value for the direct view into the laser beam is given by Eq. (20.1a): $H_{\text{MPE}} = 5 \times 10^{-3}\,\text{J/m}^2 = 5\,\text{mJ/m}^2$. This is a value very much lower than the energy density of the reference beam which is in the order of $20\,\mu\text{J/cm}^2 = 200\,\text{mJ/m}^2$. This is a factor of 40 larger than the MPE value and even reflections from the holographic plate are dangerous.

20.2.4 Multiplexing Method

Portrait holograms can also be made by multiplexing several photos which were taken of a person under different views. This process consists of three steps: photography, production of the master hologram (H1), and the reflection copy (H2).

Photography: A photographic or electronic camera is mounted on a special linear movable system (Fig. 20.3). A person or an object is positioned perpendicular to the axis of this system. The camera is moved along a straight line according to Fig. 20.3 keeping it facing forward. A series of photos is captured in regular steps. A film for slides or an electronic camera is used. Good results are obtained using eight photos [59]. However, professional systems use up to 400 photos.

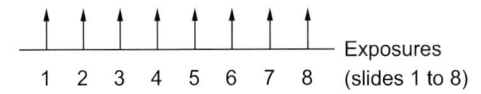

Fig. 20.3 Setup for taking eight individual photos from a person. The camera is moved linearly and remains straight ahead and is not pointed at the object. These photos are used to create a master hologram by multiplexing techniques shown in Fig. 20.4.

Master hologram (H1): A transmission hologram is made from the individual photos. In the following it is supposed that eight slides are used. The first slide is projected on a diffuser of ground glass and this image represents the object wave. The master hologram is separated into eight regions which are defined by eight slits. The width of the slits is 1/8 of the width of the holographic plate. The image of the first slide is stored in the region of the first slit according to Fig. 20.4. Then the second slide is used and the slit for the second position for exposing the second hologram. This multiplexing method is used to make eight holograms of the eight slides on the different regions of the holographic plate defined by the slit position. This is the master hologram (H1).

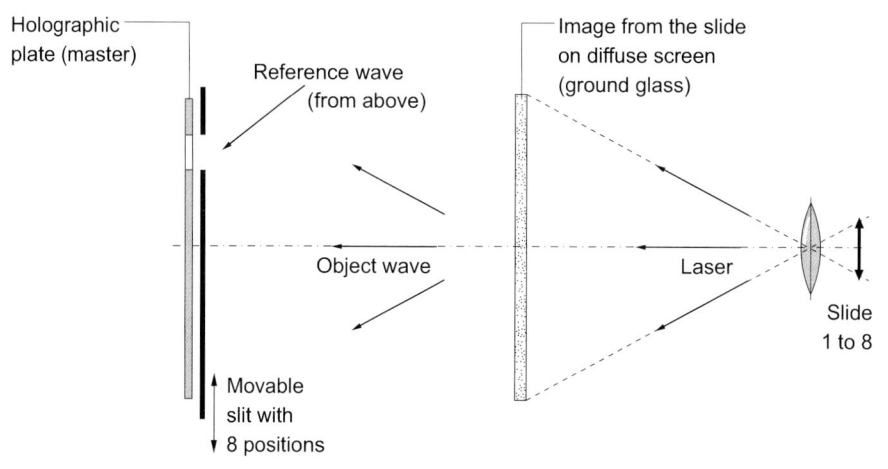

Fig. 20.4 Production of the master hologram using the images of eight slides by multiplexing. The image of each slide is stored on a different area on the hologram using a movable slit with eight positions.

Reflection copy (H2): Using this master hologram a reflection copy can be made using standard techniques as shown in Fig. 20.5a. Since the object and reference waves are incident from different directions on the holographic plate the resulting H2 is a white light reflection hologram. The reconstruction of the image from the H2 copy is shown in Fig. 20.5b. White light is incident and images of the slides one to eight are created. In addition images of the eight slits arise. Thus each slide can be seen only from a special direction. By choosing appropriate dimensions in the production steps of the holograms the distances of the slits can be correlated to the distance of the eyes as shown in Fig. 20.5b. In this figure the slit and eye distances are equal. It is also possible that the slit distance is 1/2, 1/3, 1/4 of the distance between the eyes. Thus each eye sees a different image (from a slide) and a stereoscopic image is created.

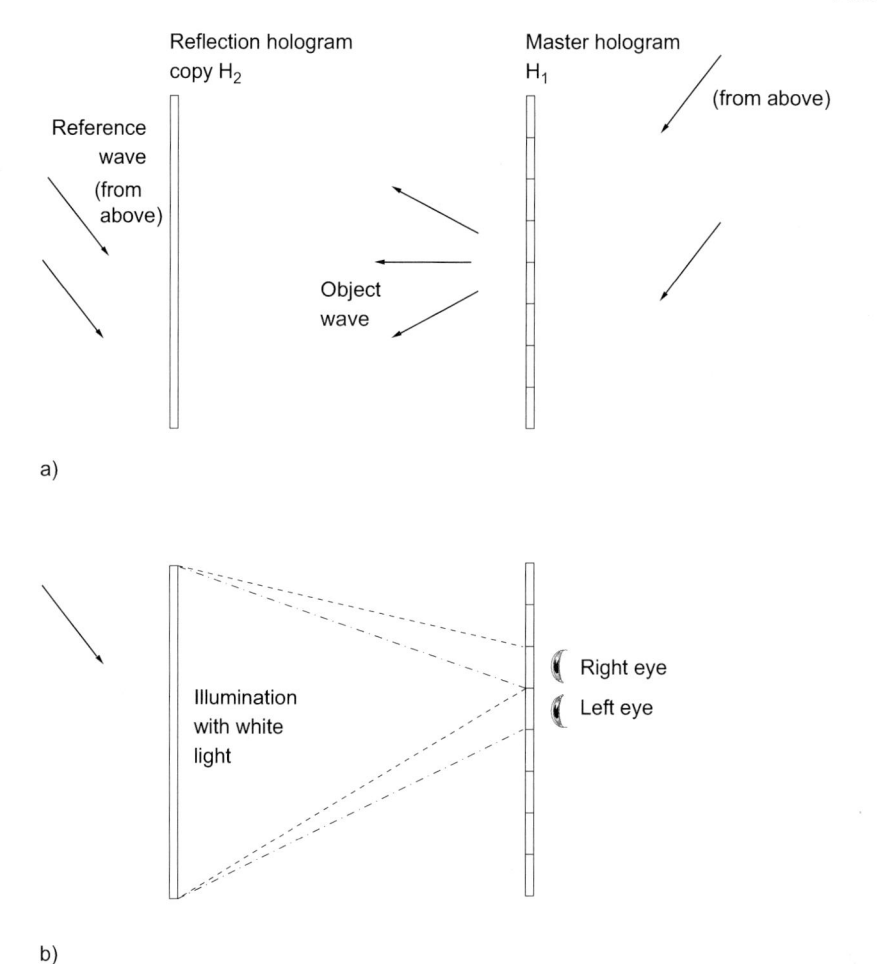

Fig. 20.5 (a) Production of a white light reflection copy (H2) in portrait holography using a multiplexed master (H1) from Fig. 20.4. (b) By white light illumination images of the eight slides and eight slits are reconstructed. Both eyes see two different images (of the slides) resulting in a stereoscopic 3D image.

Problems

Problem 20.1 Calculate the maximum permissable energy density (a) for looking directly in a laser beam and (b) for looking on a diffuser screen.

Problem 20.2 Estimate the eye save diameter of the laser beam, incident on a diffuser screen for portrait holography with a laser of (a) 1 J and (b) 4 J.

Problem 20.3 Estimate the exposure (energy density) at a distance of $x = 1$ m from a diffusing screen, which is irradiated with a laser pulse of $Q = 4$ J.

Problem 20.4 The reference beam in portrait holography is extremely dangerous. A laser beam with a radius of $w = 15$ cm ($1/e$-value) is used as a reference beam. What is the critical pulse energy Q for an eye damage?

21
Holography in Technology and Architecture

Holography is applied in many areas of technology. Some of these applications were discussed in Chapters 15 to 20. In this chapter, holography in solar energy, illumination, architecture, and aerosol detection is described.

21.1
Holography in Solar Energy

Holograms can be designed and produced for solar energy installations. Using special holographic optical elements the path of the sunlight can be modified, especially concentrated, reflected, diverted, or dispersed. For many energy applications, e.g., for power generation, broadband holograms are interesting. The solar spectrum has its maximum in the yellow–green region at about 550 nm (0.55 µm) with contributions from 300 nm (0.3 µm) to over 2000 nm (2 µm) (Fig. 21.1). Some applications select special regions of the solar spectrum, e.g., the visible, infrared or ultraviolet.

The most common holographic materials for solar energy are dichromated gelatine and photopolymers. Dichromated gelatine (DCG) can be fabricated according to the method described in the book of Saxby [39] or using recipes of other publications. Photopolymers are produced by DuPont (DuPond Holographic Materials, Wilmington. DE, USA) and Polaroid (Cambridge, MA, USA). After exposure DCG requires wet processing, whereas photopolymers require baking and ultraviolet cure. For mass production, hot stamping processes in thermoplastics are used. In contrary to application of Chapter 17, aluminization is not necessary because mostly a transmission geometry is used.

21.1.1
Photovoltaic Concentration

Holographic elements can be used to concentrate the solar radiation onto photovoltaic cells, which convert the incident radiation to electric current [60, 61]. The efficiency of these cells is about 10 to 15%. Different types of cells were developed that are sensitive in different spectral regions of the sun. Silicon

Holography: A Practical Approach. Gerhard K. Ackermann and Jürgen Eichler
Copyright © 2007 WILEY-VCH Verlag GmbH & Co. KGaA, Weinheim
ISBN: 978-3-527-40663-0

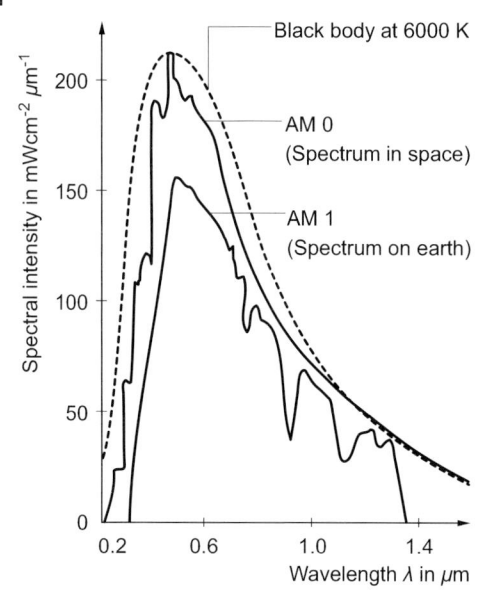

Fig. 21.1 Solar spectrum: AM1 (air mass 1): spectrum at earth with atmosphere, AM0: spectrum in space.

diodes have a maximum efficiency at about 1 µm with an upper limit at 1.1 µm, whereas GaAs is more efficient at smaller wavelengths with an upper limit of 0.89 µm. Thus, different cells should be used in the spectral solar range. In the future, quantum dot solar technology seems to be promising.

21.1.1.1 Off-axis hologram

Figure 21.2a describes the production of a hologram for concentration of the solar radiation on photovoltaic cells. A parallel and a divergent beam from the same laser are superimposed on the holographic plate. The divergent beam comes from the focal point P of a lens which is not shown. The direction of the interference fringes is given in the figure. The plate is exposed and developed. For concentrating the sunlight, the hologram is rotated by 180° around an axis in the plane of Fig. 21.2 changing the direction of the diffraction fringes. The parallel radiation from the sun is diffracted by the hologram resulting in a real pseudoscopic image of point P. Figure 21.2b shows that diffraction is stronger for longer wavelength and a spectrum is formed. This has the advantage that appropriate photovoltaic cells can be installed for different spectral regions. A typical experimental value for the intensity rise at the photocell due to holographic concentration is about 100.

If the off-axis hologram is fabricated for visible light, the distance of the interference fringes is of the order of about $d = 1$ µm. Using the simple equation

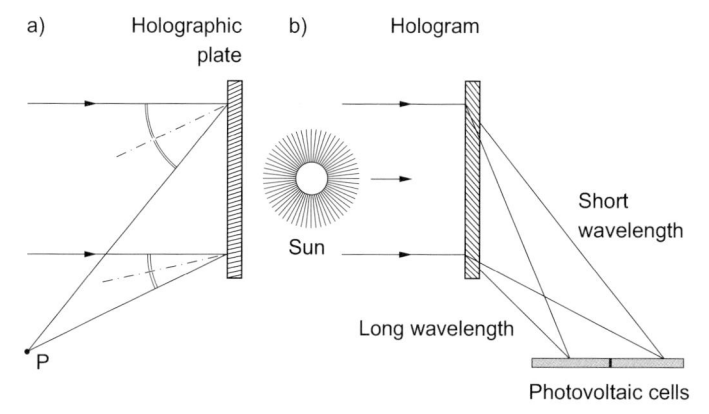

Fig. 21.2 Solar hologram of off-axis concentration of the intensity of sunlight. The lines within the hologram represent the interference fringes. (a) Production of an off-axis hologram. (b) For concentrating the sunlight, the hologram is rotated by 180°. A real (pseudoscopic) image of P is formed with strong chromatic aberration.

for a grating, the diffraction angle α is given by:

$$\sin \alpha = \frac{\lambda}{d}. \tag{21.1}$$

For a wavelength of $\lambda = 0.6$ µm, an angle of $\alpha = 37°$ results in agreement with Fig. 21.2b. For a wavelength $\lambda > d$, we get $\sin \alpha > 1$, which makes no sense. It may be concluded that infrared radiation with $\lambda > d$ passes the hologram without diffraction. This has the advantage that the photodiodes are not heated by infrared light.

21.1.1.2 In-line hologram

In-line holograms are not protected from infrared radiation and cooling may be required. The production of such a hologram is shown in Fig. 21.3a. It represents a Fresnel lens (Section 2.6). The point P is the focal point of a lens which is not shown. The hologram is rotated by 180° as described in Fig. 21.2. Illuminating the hologram by the parallel sunlight, a real image of P is created on the symmetry axis of the hologram. According to Eq. (21.1), strong chromatic aberration arises. Every wavelength has its own image resulting in a spectrum. When several diodes with different spectral ranges are arranged along the optical axis some shading effects occur. This shading may be avoided by a hologram according to Fig. 21.4. Wavelength dispersion is directed in a plane parallel to the hologram.

Any holographic system that concentrates and disperses the solar radiation requires tracking during the day. In Fig. 21.2, the sunlight must remain perpendicular to the holographic plane and the alignment of the solar cells must be precisely positioned along the spectral line.

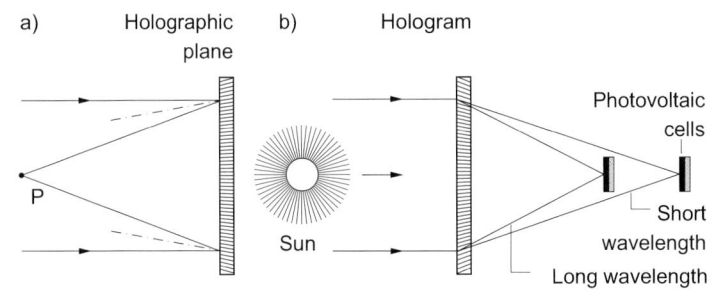

Fig. 21.3 Solar hologram for in-line concentration of the sunlight. The lines within the hologram represent the interference fringes. (a) Production of an in-line hologram. (b) For concentrating the sunlight, the hologram is rotated by 180°. A real (pseudoscopic) image of P is formed with strong chromatic aberration along the optical axis.

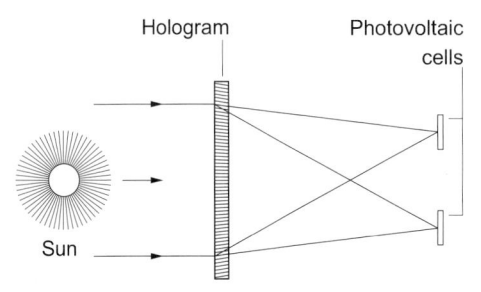

Fig. 21.4 Solar hologram for off-axis concentration of the sunlight. By chromatic aberration different wavelengths are focused in a plane perpendicular to the optical axis and different photovoltaic cells can be used.

21.2
Holography, Daylighting, and IR Blocking

21.2.1
Daylighting

Holographic daylighting systems have been developed which diffract sunlight efficiently up unto the ceiling deep in the interior of a room [60,61]. Holograms with a broad bandwidth in the visible part of the solar spectrum are used. These systems reduce both lighting and cooling costs in buildings. Figure 21.5 shows such a daylighting hologram, which consists of a transmission phase diffraction grating. The radiation of the sun is diffracted into the room. Due to dispersion on the ceiling, a spectrum of the solar light arises. The colors are seen on the ceiling only because the light is scattered in all directions. By superposition of all colors, the room is illuminated with white light. The daylighting hologram of Fig. 21.5 is similar to the hologram of Fig. 21.2 for solar energy concentration. The fabrication of daylighting holograms differs slightly from Fig. 21.2: instead of a point object P a larger area is used.

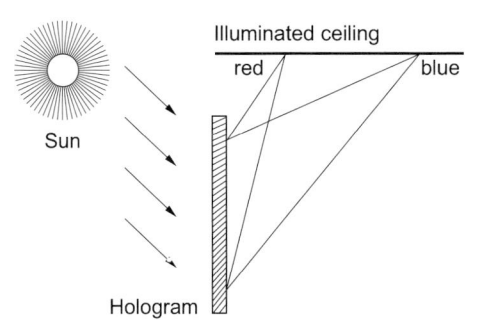

Fig. 21.5 Hologram used for daylightening of buildings. The visible part of the solar spectrum is diffracted to the ceiling of a room with chromatic aberration.

The transparent phase holograms for daylighting may be laminated directly to the glass of the windows or installed in double-paned windows. They have long lifetime and could be produced by cheap mass production. They can be constructed without mechanical tracking because they distribute the incident visible light on the ceiling from morning to evening throughout the year. Daylighting technology can result in significant savings in energy not only for lighting but also for cooling. In warm regions with a lot of sunshine, 10% of the cooling costs for air conditioning are due to heat generation by light sources.

21.2.2
Thermal Blocking

Holograms can be designed to block or reflect certain wavelength regions. In Fig. 21.6a, hologram is shown reducing the infrared radiation in buildings during summer only. The system is a reflection hologram and the fabrication is explained in Fig. 21.6a. A hologram plate or foil is matched directly with a surface mirror. It is exposed by a parallel laser beam, which is reflected by the mirror. By superposition, interference planes arise parallel to the surface of the holographic layer. A Denisjuk reflection hologram is created as described in Section 3.5, where the mirror is the object. The holographic mirror is designed for infrared radiation and a long wavelength laser should be used. In addition, postswelling processes may be used to enlarge the spacing of the interference planes to shift the hologram to the infrared part of the spectrum.

After developing the reflection hologram is laminated to a window. In summer, the sun has a large angle of incidence corresponding to the angle of the laser beam during construction of the hologram. In this case, the sun radiation is reflected according to the Bragg equation for reflection holograms:

$$2d \sin \delta = \lambda, \qquad (21.2)$$

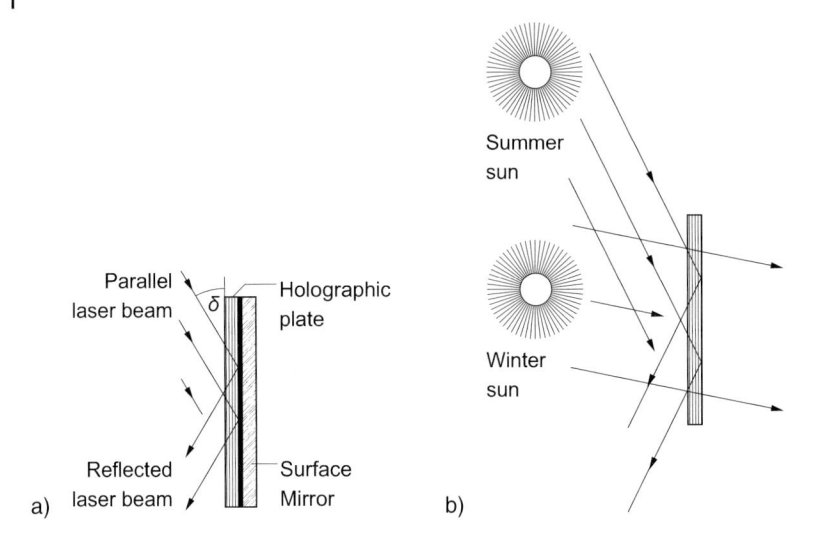

Fig. 21.6 Reflection hologram for blocking the infrared part of the solar spectrum. (a) Production of a holographic mirror (Denisjuk hologram). (b) In summer, infrared radiation is blocked by Bragg reflection from the hologram and in winter, infrared radiation passes.

where d is the spacing of the interference planes, $90°-\delta$ is the angle of incidence of the solar radiation (to the normal of the holographic plane), and λ is the wavelength. The hologram is designed for a wavelength in the infrared region and a typical angle of incidence of the summer sun. Since infrared radiation is reflected in summer heating of the rooms can be reduced and energy for cooling by air conditioners can be reduced.

In winter, the angle of incidence is smaller and δ is larger. According to Eq. (21.2), a larger wavelength region is reflected, where the infrared intensity is very low (Fig. 21.1). The major part of the visible light and near infrared is not reflected and enters the window, illuminating and heating the room.

21.2.3
Other Applications

Solar absorbers have a high efficiency when a broad wavelength range of the sun is absorbed and not reflected. Holography has been used for coating solar cells with antireflecting holographic layers. This is important because some solar cells have a high reflection coefficient due to their high refractive index. Holographic antireflecting coatings can also be applied in other areas of optics.

Other application of holography can be found in the construction of greenhouses or as special light guides. It is also possible to design holograms to select the ultraviolet part of the solar spectrum for chemistry or detoxification.

21.3
Holography in Architecture

Since the first years of holography, ideas to apply it in architecture were published and experimental work on this subject started. Many projects were performed but applications on large scale still have to be developed [62]. One of the problems is that a cheap technology to produce large holograms of several square meters is not available at the moment.

21.3.1
Documentation and Visualization

One of the obvious uses of holography is the recording of models used to previsualize new buildings. Architects usually produce models of their structures to show it to their clients. Holography can be used for three-dimensional recording of these models. The holograms can be copied, send to various locations, and easily be stored. Up to now large holograms of about 2 m^2 of this type were produced only for exhibitions and they are now found in museums. A considerable amount of work has been undertaken combining computer visualization systems with holography. In the future, computer-generated holograms may be an alternative option to models in architecture.

21.3.2
Embossed Holography

Embossed holograms are economically produced in large quantity for packing and other applications. This material is also used by artists and designers in buildings and sculptures, e.g., in indoor illuminated water cascades. The embossed material is rather sensitive and some companies incorporate it in holographic glass tiles. Embossed diffraction gratings also reflect diffused light and special directed illumination does not have to be installed.

21.3.3
Holography on Walls and Floors

Holography may be incorporated in the interior architecture of different spaces and sculptural and painted objects. Walls and floors may contain holograms or holographic optical elements. Holograms with the size of several square meters have been produced, but also smaller holograms can be used at walls and facades. At the entrance of banks and companies, large transmission holograms were installed containing the logo of the institution using color effect to draw attention. Holographic floors were installed containing reflection holograms. It is also interesting to install holograms on glass tables, e.g., in bars or restaurants.

21.3.4
HOE in Architectural Structures

Complete walls of buildings were covered with holographic grating structures. A technology similar to the dot matrix method can be applied to construct these holograms, which are composed of numerous small diffraction gratings producing special colored 3D effects. These small gratings are called holopixels which are the basis for the architectural holographic optical elements (HOE). The application of holograms on windows is described in Section 21.2.

21.4
Detection of Particles

Since the beginning of holography, this technology was applied for detection and measurement of small particles [39, 60]. These are, e.g., aerosols, droplets, or bubbles with diameters of several micrometers or more. Many technologies exist for particle size measurement and holography is only one method which can fill specific needs. The strength of this method is that particles in a relatively large volume can be imaged in one hologram. In photography, only images of particles in one plane with a small depth of field can be stored. Thus, holography offers much better possibilities to investigate dynamic situations. By producing a series of holograms, one can track in principle the path of one particle in the volume. This is not possible in photography where only one slice of the volume can be imaged. Incoherent imaging systems resolving a particle of diameter D yield the depth of field Δz

$$\Delta z = \frac{D^2}{\lambda}. \tag{21.3}$$

For a wavelength of $\lambda = 0.6$ μm and a particle diameter of $D = 3$ μm, the field of depth is only $\Delta z = 15$ μm. Therefore, it is practically impossible to store the images of particles in a large volume by photography.

21.4.1
Recording

Figure 21.7a shows a typical experimental arrangement for particle size and distribution analysis using in-line Fraunhofer holography [63]. The illumination is provided by a laser with pulses in the ns region, e.g., a Q-switched ruby laser or other solid-state laser. Pulsed radiation is necessary, because at the resolution of some μm and a velocity of about ms^{-1}, an exposure time of less than 1 μs is required. At higher velocities even shorter pulses are needed. The laser beam is focused by a lens through a spatial filter and then passes a second lens in order to make a planar wave. It illuminates parallel or divergent sample volume of, e.g., several cm^3 for particles of about 2 μm and above. The radiation scattered from the particles is the object wave and the laser beam the reference wave. A holographic plate is placed in the forward direction and a Fraunhofer hologram is created (Section 3.4). It is also possible to place a hologram plate in the backward direction which has the disadvantage that backward scattering of particle has a low intensity. A reflection hologram arises in this case.

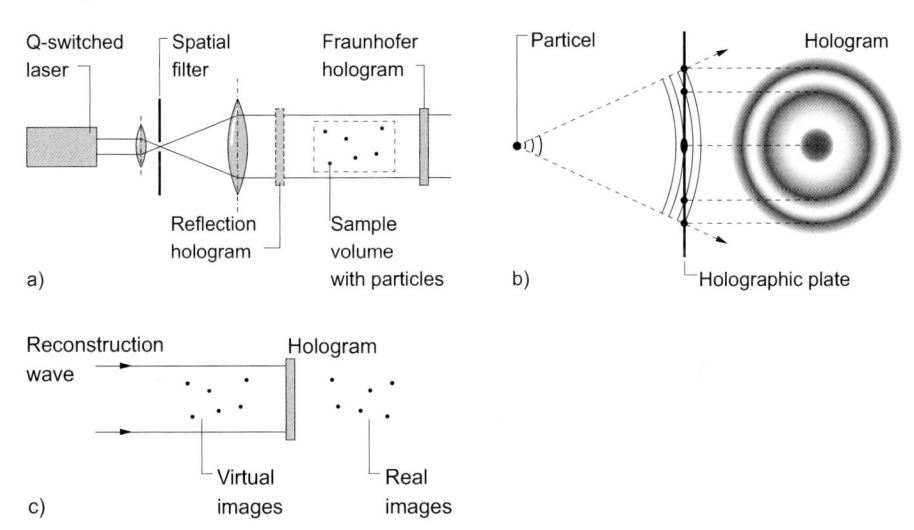

Fig. 21.7 (a) In-line Fraunhofer holographic system for particle size and distribution analysis. (b) Structure of the Fraunhofer hologram of a particle in the μm region. (c) Reconstruction: a virtual and real image is formed.

The laser radiation scattered by the particles has a complicated intensity distribution. For particles with dimension in the region of the wavelength, Mie scattering occurs with strong forward scattering. For very small particles, a nearly spherical wave arises. This case is given in Fig. 21.7b, where the superposition of a spherical object wave and a plane reference wave is shown. The hologram is represented by a Fresnel lens (Section 2.6). The Fraunhofer hologram of a sample volume with particles is the superposition of structures similar to Fig. 21.7b. At most 20% of the recording beam cross section may encounter particles.

21.4.2
Reconstruction

For measurement of the particle size, the Fraunhofer hologram is illuminated by a laser beam (Fig. 21.7c). Virtual images of the particle are formed at the original position of the particles if the geometry of recording at reconstruction is maintained. In addition, in the forward direction a real image arises. A screen can be used to observe the real image in different planes. It is more convenient to use a TV system or a digital camera to scan the virtual or real image.

It can be seen from Fig. 21.7b that the radius of the nearly spherical object wave increases with distance from the object particle. In consequence, the distance of the spherical interference fringes in the Fraunhofer hologram becomes larger too. It is possible to choose a distance where the pixel size of a digital camera is able to resolve the interference fringes. Thus, digital or electronic holography may be applied for particle detection and measurement.

Problems

Problem 21.1 An interference grating ($d = 0.9$ µm) is used in transmission to separate visible and infrared radiation. Calculate the angle of diffraction α for the first order for wavelength of $\lambda = 0.4, 0.7, 0.8, 0.9$, and > 0.9 µm.

Problem 21.2 A hologram for inline concentration of solar light is produced with a wavelength of $\lambda = 0.532$ µm according to Fig. 21.3a. The distance of the point P in Fig. 21.3a is 40 cm. Calculate the focal length f' in Fig. 21.3b for $\lambda = 0.4, 0.532$, and 0.7 µm. Which laser is used?

Problem 21.3 Calculate the radius r of the first interference ring created in inline Fraunhofer particle detection according to Fig. 21.3 at a distance of $z = 10$ cm from the particle. A He–Ne laser is used ($\lambda = 0.632$ µm).

Problem 21.4 How large is the depth of field for the conventional particle detection (He–Ne laser with 632 nm) for a particle size of $D = 1$ µm and $D = 10$ µm? Compare it with the holographic method.

Appendix A
Wave Functions and Complex Numbers

A.1
Complex Numbers

A complex number is defined as

$$\mathbf{z} = a + ib. \tag{A.1}$$

Here a and b are real numbers, where a is the real part and b the imaginary part of $\mathbf{z}$ and "i" being the imaginary unit. It is

$$i = \sqrt{-1}. \tag{A.2}$$

The number $\mathbf{z}$ is drawn within a two-dimensional system, given in Fig. A.1

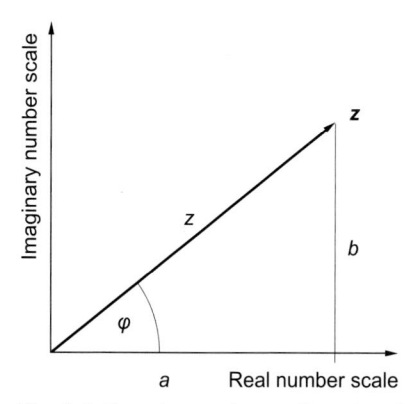

Fig. A.1 Complex number z with real part a and imaginary part b. Shown is the length z and the angle ϕ between z and the "real number scale," too.

There are two more definitions of $\mathbf{z}$. Given the angle ϕ between $\mathbf{z}$ and the real number scale and the length of $\mathbf{z}$, $z = |\mathbf{z}|$ then from Fig. A.1 it is

$$\mathbf{z} = z \cdot (\cos\phi + i \cdot \sin\phi). \tag{A.3}$$

Holography: A Practical Approach. Gerhard K. Ackermann and Jürgen Eichler
Copyright © 2007 WILEY-VCH Verlag GmbH & Co. KGaA, Weinheim
ISBN: 978-3-527-40663-0

With Euler's theorem

$$e^{i\phi} = \cos\phi + i\cdot\sin\phi$$

it is

$$\mathbf{z} = z \cdot e^{i\cdot\phi}. \tag{A.4}$$

A.2
Wave Functions, Sine and Cosine Waves

A wave is a periodic function in time and space. The time dependence is discussed, first. If in Fig. A.2 the point P spins counter clockwise round M with a constant velocity, then the projection of point P, P_0 describes a harmonic oscillation as shown.

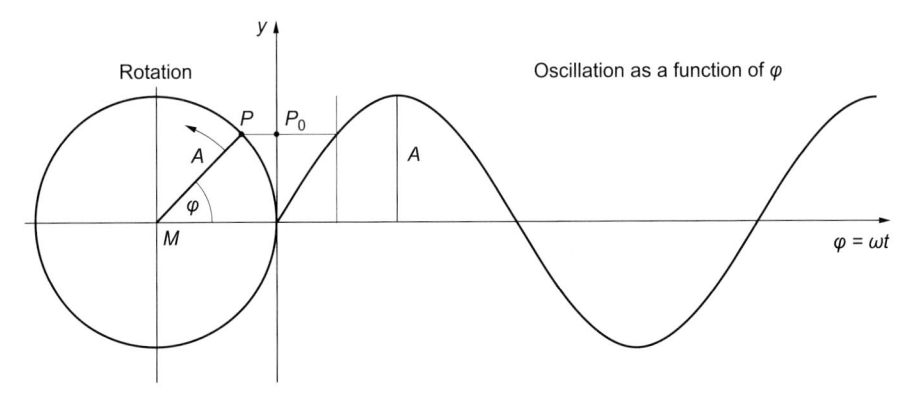

Fig. A.2 Representation of an oscillation $\phi = \omega t$.

This oscillation is given by

$$y = A\cdot\sin\phi, \tag{A.5}$$

where A is the amplitude and ϕ is the angel of rotation. ϕ can be expressed as a function of time t and the angular frequency of the rotation ω. It is

$$\phi = \omega\cdot t, \tag{A.6}$$

with $\omega = 2\cdot\pi\cdot f$ and $f = 1/T$; f is the frequency and T is the time for one rotation.

The frequency ν is defined as

$$\nu = \frac{1}{T}$$

A.2 Wave Functions, Sine and Cosine Waves

and the angular frequency as $\omega = 2\pi \cdot \nu$. Inserting Eq. (A.6) in Eq. (A.5) it is

$$y = A \cdot \sin(\omega t). \tag{A.7}$$

As mentioned above, a wave is also a function of space coordinates. Here one space coordinate, x, is considered, only. For $t = $ const. the periodicity of the function is determined by λ, the wavelength. λ is the distance between two maxima or minima of the given wave y. It follows

$$y = A \cdot \sin\left(2\pi \frac{x}{\lambda}\right), \tag{A.8}$$

λ being the wavelength. With the abbreviation

$$k = \frac{2\pi}{\lambda}$$

Eq. (A.8) can be written as

$$y = A \cdot \sin(kx), \tag{A.9}$$

where k is the "angular wave number" or the "wavelength constant" or "propagation constant."

If the instantaneous amplitude is not zero for $t = 0$, a phase angle has to be added or subtracted in Eq. (A.7)

$$y = A \cdot \sin(\omega t - \phi_0). \tag{A.10}$$

Subtracting ϕ_0 in Eq. (A.10) is equal to shifting the sine function to the right. Now the phase angle can be interpreted as another point x_1 at the same time t. Then we get with $\phi_0 = kx_1$

$$y = A \cdot \sin(\omega t - kx_1). \tag{A.11}$$

Equation (A.11) holds for all values of t and x_1. This equation describes a wave moving in the positive x direction. Equation (A.11) represents a wave function. The general wave function contains another phase constant, because the instantaneous amplitude may not be zero for $t = 0$ and $x = 0$. It follows

$$y = A \cdot \sin(\omega t - kx + \alpha), \tag{A.12}$$

α may be positive or negative. Equation (A.12) is equivalent with the formula

$$y = A \cdot \cos(\omega t - kx + \beta). \tag{A.13a}$$

In Eq. (A.13a) describing the same wave as Eq. (A.12), the phase constant β of course is different by $\pi/2$ from α in Eq. (A.12). The wave equations given here are special for two reasons. One spatial coordinate is shown, only. A wave propagating is evolving in three dimensions. Then instead of x the radius vector **r** is used. Very often the propagation constant k is taken as a vector **k**. The length of **k** is given above as $(2\pi/\lambda)$, the direction of **k** points into the direction of wave propagation. Then for Eq. (A.13a) it is

$$y = A \cdot \cos(\omega t - \mathbf{kr} + \alpha), \tag{A.13b}$$

kr being the scalar product of the vectors **k** and **r**; α is a new constant phase factor. All equations up to (A.13a) are linear polarized wave functions. The oscillations are within a fixed plane, the plane of the paper. Light from a laser or a lamp may oscillate in a plane perpendicular to the propagation direction and may also rotate within this plane during propagation (elliptical or circular polarized radiation). A is a time constant, but may be varying with the spatial coordinates. So it is $A = A(x, y, z)$.

A.3
Wave Functions, Exponential Representation

In order to derive the exponential description of wave functions, Eq. (A.12) is taken. If kx and α are combined into one phase ϕ, it is

$$y = A \cdot \cos(\omega t - \phi). \tag{A.14}$$

From basic math one can derive:

$$y = A \cdot \cos\phi \cdot \cos\omega t + A \cdot \sin\phi \cdot \sin\omega t,$$

or

$$y = A_1 \cdot \cos\omega t + A_2 \cdot \sin\omega t. \tag{A.15}$$

The amplitudes A_1 and A_2 contain the same information as A and ϕ in Eq. (A.14). A_1 and A_2 behave like vectors or like the real and imaginary parts of a complex number. They are called phasors. Summing up A_1 and A_2 – as given below – the original values of A and ϕ result.

Using Eqs. (A.1) and (A.4) a new formulation of wave functions can be found. It is (see Fig. A.3):

$$\mathbf{A} = A_1 + i \cdot A_2,$$

and

$$\mathbf{A} = A \cdot \cos\phi + i \cdot A \cdot \sin\phi. \tag{A.16}$$

Applying Euler's law

$$\mathbf{A} = A \cdot e^{i \cdot \phi}, \tag{A.17}$$

A.3 Wave Functions, Exponential Representation

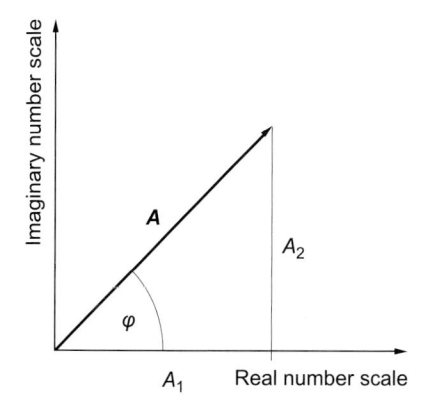

Fig. A.3 Addition of phasors.

where A is called the complex amplitude. In most cases in holography the time dependence is not significant. The complex amplitudes are used always in holography. Very often – not quite correct – those complex amplitudes are just called wave function. In this book Eq. (A.17) is very often used instead of the complete wave function.

In completing this introduction into wave functions Eq. (A.17) can be expanded by inserting the time dependence. In Eq. (A.15) the trigonometric functions can be written as

$$\cos \omega t = \frac{1}{2}\left(e^{i\omega t} + e^{-i\omega t}\right)$$

and

$$\sin \omega t = \frac{1}{2i}\left(e^{i\omega t} - e^{-i\omega t}\right).$$

Inserting these expressions into Eq. (A.15) and sorting the expressions with respect to different exponential functions yields

$$\mathbf{y} = \frac{1}{2}(A_1 - iA_2) \cdot e^{i\omega t} + \frac{1}{2}(A_1 + iA_2) \cdot e^{-i\omega t}$$

or

$$\mathbf{y} = \frac{1}{2}A^* e^{i\omega t} + \frac{1}{2}A e^{-i\omega t}. \tag{A.18}$$

Equation (A.18) describes a sum of a complex number and their conjugate. Drawing the complex numbers for any given time t within a coordinate system like Fig. A.1, results into two imaginary parts, which always cancel out each other and two real parts, which sum up to give a cosine wave. Therefore, it is convenient to define a wave function taking just the double of one part of Eq. (A.18)

$$\mathbf{y} = A \cdot e^{-i(\omega t - \mathbf{kr} + \alpha)}. \tag{A.19}$$

Here A is the real amplitude of the wave. The exponential form of wave functions is mathematically easier to handle than sine or cosine functions. For example the square of the function, often used within holography to calculate the intensity is

$$I = \mathbf{y} \cdot \mathbf{y}^*,$$

$\mathbf{y}^*$ being the conjugate complex of $\mathbf{y}$. It follows

$$I = \left(A \cdot e^{i(\omega \cdot t + k \cdot x + \alpha)}\right) \cdot \left(A \cdot e^{-i(\omega \cdot t + k \cdot x + \alpha)}\right)$$

$$I = A^2.$$

In order to get back to the sine or cosine form of the wave function, one takes, e.g., the real part "Re" of Eq. (A.19), which then with Euler's equation gives the cosine function,

$$y = A \cdot \mathrm{Re}\left(e^{-i(\omega t - kr + \alpha)}\right).$$

Appendix B
Bragg Diffraction

Bragg diffraction occurs when light is passing through a transparent uniformly layered object. A well-known example of Bragg diffraction are the Laue diagrams, which in 1911 disclosed the nature of X-rays and the structure of single crystalline objects. The observed pattern of the scattered light stored in a photographic plate could be well described with the Bragg formalism.

In holography Bragg diffraction is observed when light is passing through the exposed, developed, and bleached transparent emulsion of a holographic plate with layered alternating refractive index. An example is given in Fig. B.1, which is a transmission phase hologram made of two plane waves.

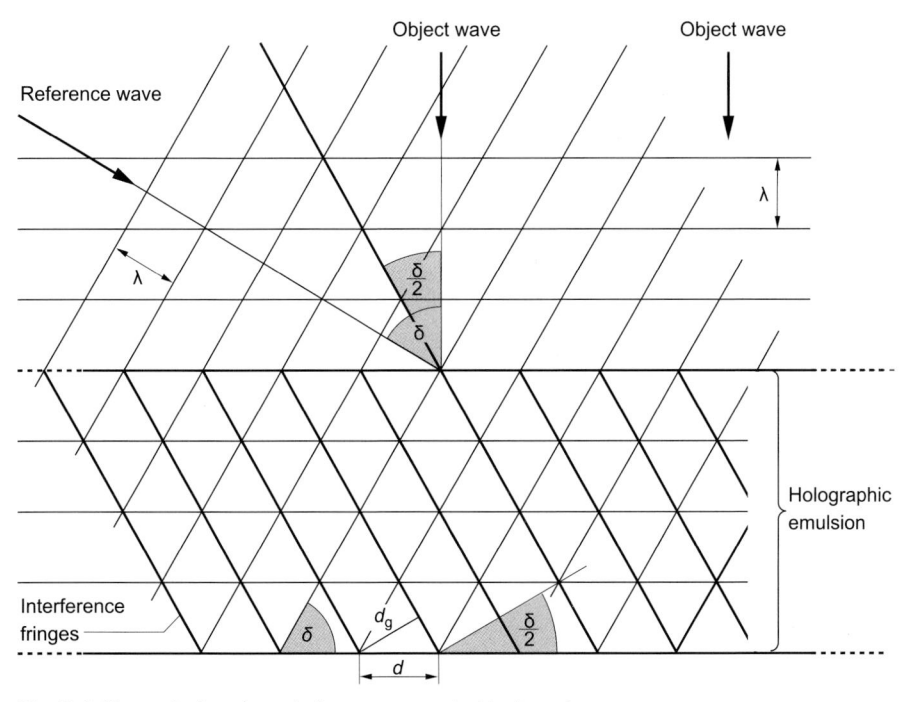

Fig. B.1 Transmission phase hologram generated by two plane waves of coherent light perpendicular to the emulsion surface and under an angle δ with respect to the perpendicular direction.

Holography: A Practical Approach. Gerhard K. Ackermann and Jürgen Eichler
Copyright © 2007 WILEY-VCH Verlag GmbH & Co. KGaA, Weinheim
ISBN: 978-3-527-40663-0

The transmission hologram consists of fringes parallel to the bisector of the reference beam, and the object beam, respectively. The fringes are of higher refractive index compared with the surrounding material. If the reference beam hits the developed hologram under the angle used during exposure, then the object beam is reconstructed (Snellius' law is neglected at the interfaces in the figures given here, because the effect, when the light is entering the hologram is cancelled out, when the same light is exiting the hologram).

The planes of higher refraction index are semitransparent mirrors. Part of the wave is reflected; part is transmitted and will hit the next plane of higher refractive index. If one do not use the angle of exposure, the reconstructed object beam is of low intensity or the reconstruction fails at all. The reason is that the so-called Bragg condition is not obeyed. Of course, in this simple experiment one can exchange object and reference beam during the reconstruction process. The plane object wave reconstructs the plane reference wave as well as vice versa. Very often one calculates the direction of this reconstructed beam for a reconstructing beam meeting the hologram perpendicular to the surface by just using the grating formula (Chapter 8, experiment No. 10), only:

$$\sin \delta = \frac{\lambda}{d}, \tag{B.1}$$

here λ is the wavelength of the laser light, d is the grating constant, and δ is the diffraction angle (Figs. A2.1 and A2.2). If this is a true description, then the conjugated beam (first order of diffraction) should be visible, too. But that is not the case. Here the Bragg condition comes into play, named after William Bragg and his son Lawrence, who first dealt with the general problem and won the Nobel prize in 1915.

As shown in Fig. B.2, the path differences of different partitions of the reconstructing wave meeting different semitransparent planes within the hologram must be λ or a multiple of λ in order to be in phase with the other partitions. Then a maximum intensity will be reached. The path differences depend on the spacing of the fringes. After Fig. B.1 it is

$$d_g = d \cdot \cos \frac{\delta}{2}. \tag{B.2}$$

Here d_g is the spacing of the higher index of refraction planes, d is the distance of the fringes at the surface of the holographic plate. In a thin hologram d would be the grating constant and δ is the reconstruction angle. Now it is from basic mathematics

$$\sin \delta = 2 \sin \frac{\delta}{2} \cos \frac{\delta}{2}. \tag{B.3}$$

Inserting Eqs. (B.3) and (B.2) into Eq. (B.1) yields

$$2 \sin \frac{\delta}{2} \frac{d_g}{d} = \frac{\lambda}{d},$$

Reconstruction wave

Holographic
emulsion

Reconstructed wave

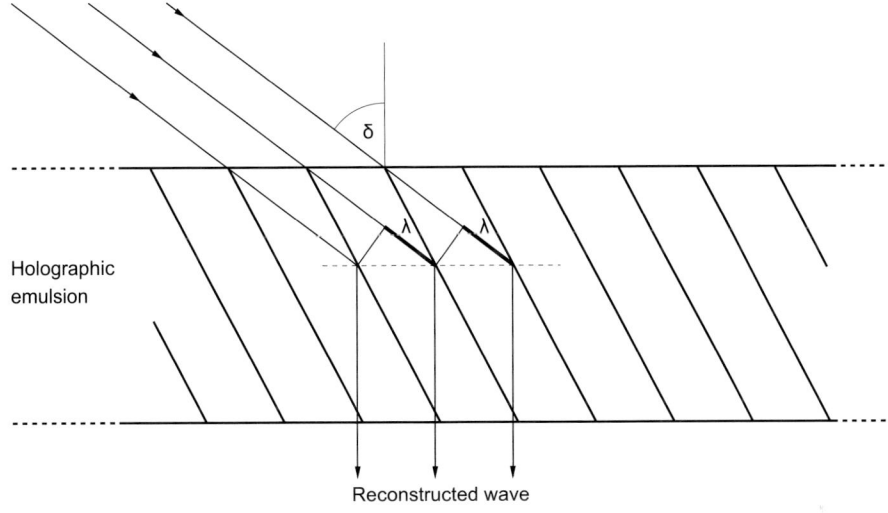

Fig. B.2 Reconstruction with a plane wave, which hits the emulsion under an angle δ.

or

$$2d_g \sin \frac{\delta}{2} = \lambda. \tag{B.4}$$

Equation (B.4) is the so-called Bragg condition. Maximum light within the reconstructed object wave is given; if this condition is obeyed. The conjugated wave does not fulfill the Bragg condition and will not be reconstructed. Now it is quite clear, that the grating equation (B.1) gives the right result for the reconstructed object wave only.

If the angle between the reference wave and object wave is increased to 90° (grazing incidence), then the Bragg planes are at 45°. Now with a fixed δ, λ is fixed too. Volume transmission holograms are wavelength sensitive.

If δ is increased further up to 180°, then we have more Bragg planes within the emulsion and those reflection holograms are even more wavelength sensitive. Therefore, these holograms can be reconstructed using the white light.

Appendix C
Fourier Transform and Fourier Hologram

C.1
Fourier Series

In order to understand Fourier holograms and Fourier transform, Fourier series are introduced, first. To simplify the arguments, one variable x is used only.

A periodic function given in Fig. C.1 can be represented by an infinite series of sine and cosine functions, respectively:

$$f(x) = \frac{a_0}{2} + \sum_{n=1}^{\infty} a_n \cdot \cos\left(\frac{2\pi n x}{D}\right) + \sum_{n=1}^{\infty} b_n \cdot \sin\left(\frac{2\pi n x}{D}\right). \tag{C.1}$$

The arguments within the trigonometric functions are angles in radian built from the multiple of x over D, $n(x/D)$, multiplied by 2π. D is the period

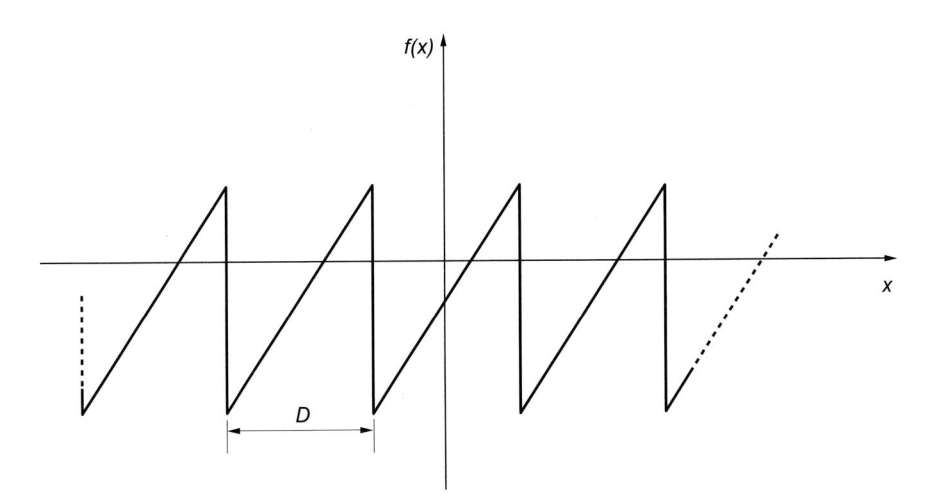

Fig. C.1 Periodic function $f(x)$. D: Period of periodic function.

Holography: A Practical Approach. Gerhard K. Ackermann and Jürgen Eichler
Copyright © 2007 WILEY-VCH Verlag GmbH & Co. KGaA, Weinheim
ISBN: 978-3-527-40663-0

of the periodic function $f(x)$. The running number n counts the multiple of $(2\pi x/D)$.

The coefficients a_0, a_n, and b_n, attached to these functions describe the strength or amplitude of a given trigonometric function within the series, which represent the given periodic function. They are determined by the formulas

$$a_0 = \int_D f(x)\,dx,$$

$$a_n = \int_D f(x) \cdot \cos\left(\frac{2\pi nx}{D}\right) dx, \quad \text{and} \tag{C.2}$$

$$b_n = \int_D f(x) \cdot \sin\left(\frac{2\pi nx}{D}\right) dx.$$

Where D is the period of $f(x)$ as shown in Fig. C.1. The integral has to be calculated over one period D. The formulas of Eq. (C.2) are derived by integrating Eq. (C.1) for a given a_n or b_n. D could be the grating constant of a holographic grating. In this case $u_n = (n/D)$ are the spatial frequencies. x and u are called the conjugated parameters.

For a continuous grating, given in Fig. C.2, with grating constant D, width of a slit a and a constant illumination h the Fourier series can be calculated from Eq. (C.1).

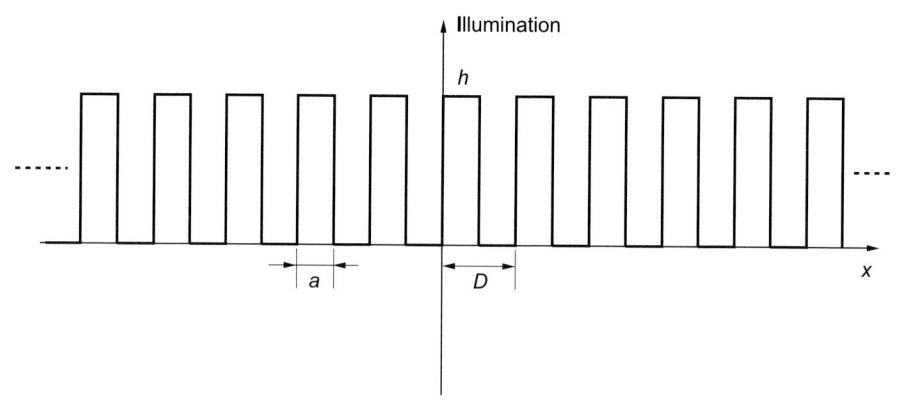

Fig. C.2 Continuous grating; D: grating constant, a: width of grating slit; $f(x)$ is the transmission of the grating (constant illumination value h).

First it is obvious from Fig. C.2 that a grating is an even function and all b_n are zero. Then it is

$$f(x) = \frac{a_0}{2} + \sum_{n=1}^{\infty} a_n \cdot \cos\left(\frac{2\pi nx}{D}\right). \tag{C.1a}$$

For the coefficients a_n it is

$$a_n = \int_0^{(a/2)} \frac{h}{D} \cdot \cos\left(\frac{2\pi n x}{D}\right) dx + \int_{(D-a/2)}^{D} \frac{h}{D} \cdot \cos\left(\frac{2\pi n x}{D}\right) dx,$$

which is equivalent with the integral

$$a_n = \frac{2 \cdot h}{D} \int_0^{(a/2)} \cos\left(\frac{2\pi n x}{D}\right) dx.$$

The result is the very well-known sinc function

$$\frac{a_n}{2} = \frac{ha}{D} \cdot \frac{\sin\left(\frac{\pi n a}{D}\right)}{\frac{\pi n a}{D}} = \frac{ha}{D} \cdot \mathrm{sinc}\left(\frac{\pi n a}{D}\right). \tag{C.3}$$

In Fig. C.3 the coefficients a_n are given as a function of spatial frequency (n/D).

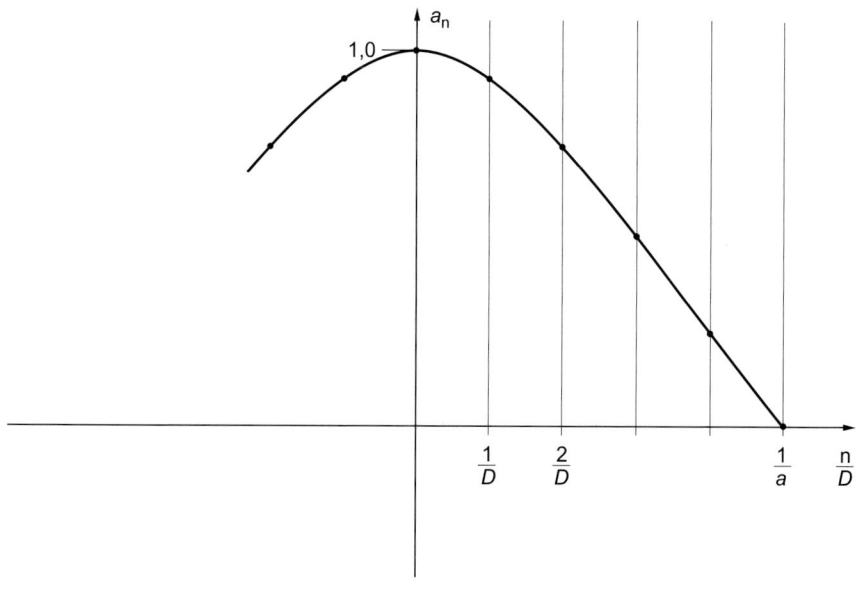

Fig. C.3 Coefficients a_n as a function of spatial frequency in relative units. In Eq. (C.3) the constant value $(2ha/D)$ was normalized to be 1 and it was taken $a = 0.2D$.

The square of the coefficients a_n give the intensity of the light within different diffraction orders.

C.2
Exponential Form of Fourier Series

Applying Euler's law, Eq. (C.1) can be written as

$$f(x) = \frac{a_0}{2} + \frac{1}{2}\sum_{n=1}^{\infty}(a_n + ib_n)\cdot e^{\frac{2\pi i n x}{D}} + \frac{1}{2}\sum_{n=1}^{\infty}(a_n - ib_n)\cdot e^{-\frac{2\pi i n x}{D}}.$$

The second summation sums up the complex conjugates of the first summation. If the second summation is made to run from $n = -1$ to $n = -\infty$, then the minus sign of the second summation vanishes. If in addition the sum of the coefficients is abbreviated by A_n and A_n^*, then it follows for $f(x)$

$$f(x) = \sum_{n=-\infty}^{\infty} A_n \cdot e^{\frac{2\pi i n x}{D}}, \tag{C.4}$$

and it is $A_n = \frac{1}{2}\cdot(a_n + i\cdot b_n)$.

For the negative values of n the coefficients are the conjugated complex of A_n. The coefficients are found by integrating Eq. (C.4):

$$A_n = \frac{1}{D}\cdot \int_D f(x)\cdot e^{-\frac{2\pi i n x}{D}}dx. \tag{C.5}$$

C.3
Fourier Transform

The last two equations are very close to the Fourier transform formulas. The important difference is that the Fourier transform is applicable to nonperiodic functions as well.

First, consider again the result of the grating, given above. Equations (C.1) and (C.2) describe the problem for an even function ($b_n = 0$). If the grating constant increases then the distance between maxima and minima or spatial frequencies (n/D) decreases. If the grating constant increases to infinity, there is one slit left only and all spatial frequencies are allowed; n and D lose their significance, the summation in Eq. (C.1) now is an integral. The new integrals are

$$f(x) = \int_{-\infty}^{\infty} F(u)\cos(2\pi u x)\,du \quad \text{and} \quad F(u) = \int_{-\infty}^{\infty} f(x)\cos(2\pi u x)\,dx, \tag{C.6}$$

$f(x)$ and $F(u)$ being the Fourier transforms of each other. Because the grating problem deals with a real and even function, Eq. (C.6) is called the cosine Fourier transform.

When $f(x)$ is not even or odd one would have to consider sine and cosine functions. Instead the exponential function is more convenient. In addition, it is applicable to real and complex functions, too. From Eqs. (C.4) and (C.5) it is

$$f(x) = \int_{-\infty}^{\infty} F(u) \cdot e^{2\pi i u x} du, \quad \text{and} \quad F(u) = \int_{-\infty}^{\infty} f(x) \cdot e^{-2\pi i u x} dx. \qquad (C.7)$$

For Fourier holograms Eq. (C.7) is needed for two variables. The variables are x and y and u and v, respectively. The Fourier transform is

$$f(x,y) = \int_{x=-\infty}^{x=\infty} \int_{y=-\infty}^{x=\infty} F(u,v) \cdot e^{2\pi i (ux+vy)} du dv, \quad \text{and}$$

$$F(u,v) = \int_{x=-\infty}^{x=\infty} \int_{y=-\infty}^{y=\infty} f(x,y) \cdot e^{-2\pi i (ux+vy)ux} dx dy \qquad (C.7a)$$

In the grating example x and u are the spatial coordinate and the spatial frequency, respectively. Equation (C.7) is called the Fourier transform and the variables conjugate parameters. $f(x)$ and $F(u)$ are a Fourier transform pair.

The Fourier transform of a single slit is calculated with Eq. (C.7), again. If $f(x)$ represents a single slit, then it is

$$f(x) = h, \quad -\frac{a}{2} \leq x \leq \frac{a}{2}. \qquad (C.8)$$

The Fourier transform is

$$F(u) = h \cdot \int_{-\frac{a}{2}}^{\frac{a}{2}} e^{-2\pi i u x} dx.$$

The integration is done easily and delivers the well-known sinc function, again:

$$F(u) = h \cdot a \cdot \sin c (\pi u a).$$

C.4
Convolution and Correlation

Two other integral formulas are important for solving optical problems. If $f(x)$ and $g(x)$ are two functions, the convolution integral is defined as

$$h(x) = \int_{-\infty}^{\infty} f(x') \cdot g(x - x') dx' = f(x) \otimes g(x). \qquad (C.9)$$

The convolution is best described as the result of the original function $f(x)$ – e.g., light emitted of a lamp or a star – which is detected with a device of limited resolution $g(x)$ in time or space. Then the original function is to some extent blurred because of the limited resolution. The detecting function is centered on x; it is displaced by $x\prime$ and reflected while the original function is at $x\prime$. The integral is taken over all possible values of $x\prime$. The integral $h(x)$ describes the convolution. The detecting function $g(x)$ could be the resolution of a multiplier or could be described by the diffraction of a point like star within an astronomical telescope.

The cross-correlation function is given by

$$c(x) = \int_{-\infty}^{\infty} f(x') \cdot g(x'-x) dx' = f(x) (\bullet) g(x), \tag{C.10a}$$

and for complex functions

$$c(x) = \int_{-\infty}^{\infty} \mathbf{f}(x') \cdot \mathbf{g}^*(x'-x) dx' = \mathbf{f}(x) (\bullet) \mathbf{g}^*(x). \tag{C.10b}$$

The autocorrelation function is given by

$$c(x) = \int_{-\infty}^{\infty} f(x') \cdot f(x'-x) dx' = f(x) (\bullet) f(x), \tag{C.11a}$$

and for complex functions

$$c(x) = \int_{-\infty}^{\infty} \mathbf{f}(x') \cdot \mathbf{f}^*(x'-x) dx' = \mathbf{f}(x) (\bullet) \mathbf{f}^*(x). \tag{C.11b}$$

In both cases x is the shift variable and $x\prime$ the integration parameter. The above equations prove the similarity of two functions or for the correlation of a function with itself. These functions are very useful to describe the reconstruction process of a hologram. Equation (C.11a) calculates the halo, observed during reconstruction. In this case we have to take the object function $o(x,y)$. The integral equation (C.11b) has to be written for two coordinates.

In order to see the effect of autocorrelation a rectangle is given in Fig. C.4

The rectangle is given and the same rectangle displaced by a and three quarters of a, if a is the base of the rectangle. The value of the autocorrelation integral for the displacement is the overlap area. The result of the autocorrelation for all possible shifts is the triangle; given in Fig. C.4 too. It can be seen that the autocorrelation function is by no means similar to the original function. Often it is necessary to normalize the output of Eqs. (C.10a,b) and (C.11a,b). Then

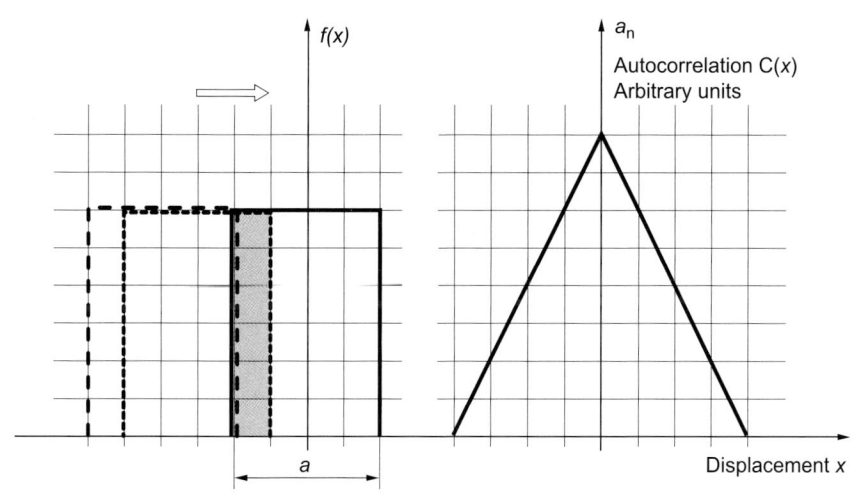

Fig. C.4 Autocorrelation for a given rectangle function, $f(x)$ = constant for $(-a/2) \leqslant x \leqslant (a/2)$. The baseline of the rectangle is a. The baseline of the autocorrelation function is $2a$.

the result of the integrations is divided by the integral for zero shift. From the technique of calculation it is quite obvious (Fig. C.4) that the baseline of the autocorrelation function is two times the baseline of the given function.

Thinking of the holographic reconstruction process it is obvious that the halo, observed around the zeroth order is nothing else but the autocorrelation function of $o(x,y)$. One has to consider that during exposure any part of the object wave is superimposed by any other part of the object wave. The result is the observed halo. The spread of the halo is two times the spread of $o(x,y)$ in accordance with the autocorrelation process.

Solutions to the Problems

Chapter 1

Problem 1.1 The two essentials are the amplitude A and the phase ϕ. A (plane) wave is given as $y(\mathbf{r},t) = Ae^{i\phi(\mathbf{r},t)}$, $\mathbf{r}$ being the radius vector and t the time (see Appendix A).

Problem 1.2 Within a photographical exposure the amplitude A of an object is stored, only. The phase that contains the three-dimensional geometry of the object is discarded due to the long exposure time compared with the cycle time of the wave.

Problem 1.3 The phase of the object wave is preserved because of an interference effect. The coherent reference and object wave form an interference pattern, which is independent of time, but which depend on the structure of the scattered object wave. So, the optical principle of exposure is interference. During reconstruction the optical phenomenon is diffraction, only. Therefore no coherent light is necessary for reconstruction.

Problem 1.4 No. With only one wave, there is no time independent interference pattern. The phase will be discarded with incoherent and coherent light. (see Problem 1.2)

Chapter 2

Problem 2.1 The visibility is given as $V = \frac{I_{max}-I_{min}}{I_{max}+I_{min}}$. It is $I_{max} = (A_o + A_r)^2 = 9$ and $I_{min} = 1$. It follows $V = 0.80$ or 80%.

Problem 2.2 $u(x,y) = o(x,y) \cdot t(x,y)$. Now, applying Eq. (2.22a) for $t(x,y)$, it follows:

$$u(x,y) = \left(t_0 + \beta\tau \cdot r^2\right) \cdot o(x,y)$$
$$+ \beta\tau o^2(x,y) \cdot o(x,y)$$
$$+ \beta\tau o^2 r(x,y)^* o(x,y) \cdot o(x,y)$$
$$+ \beta\tau o^2 \cdot r(x,y)$$

The last term shows the reconstructed reference wave.

Problem 2.3 If σ_0 is the maximum spatial frequency of the object wave, then Eq. (2.30) shows that the halo around the zeroth order (σ_r) is $\pm 2\sigma_0$. The object wave around $\sigma = 0$ has an extension of $\pm\sigma_0$. The real image at $\sigma = 2\sigma_r$ has the same extension. In order to prevent overlapping of object and reconstruction wave, if follows $\sigma_r > 3\sigma_0$.

The extension of the halo of $2\sigma_0$ can be verified with the help of the autocorrelation function in Appendix C: Assuming for simplicity that the wave is a rectangular function within the spatial frequency regime, then as shown in the Appendix the autocorrelation function has a base of $\pm 2\sigma_0$ if the base of the function is $\pm\sigma_0$.

Problem 2.4 A conjugated wave is described by the complex conjugate of a given wave function. The wave travels in the opposite direction and has the opposite curvature compared with the original wave. (Convergent in contrast to divergent or vice versa)

Problem 2.5 From Eq. (2.36) it follows $2r_k \frac{dr_k}{dk} = 2z_0\lambda + 2k\lambda^2$. Taking the differentials as differences it follows for $\Delta k = 1$ Eq. (2.37).

Problem 2.6 $r_1 = 0.12$ mm; $r_{100} = 3.5$ mm.

Chapter 3

Problem 3.1 The reconstruction wave and the object waves travel in the same direction. Observation of the reconstructed object wave is therefore disturbed. In addition the virtual and the real images travel in the same direction.

Problem 3.2 During exposure of a transmission hologram the object wave and the reference wave impinge the holographic plate from the same side. For a reflection hologram the reference wave and the object wave impinge the plate from opposite sides.

In reconstruction the object waves are reconstructed and these waves travel in the same direction as during exposure. To see the image of the object one has to look more or less toward the reconstruction wave and through the hologram for a transmission hologram and in the direction of the reconstruction wave for a reflection hologram. In both cases the virtual image lies behind the hologram.

Problem 3.3 For transmission holograms the interference fringes or planes are positioned almost perpendicular to the emulsion surface, depending on the angle between the reference and object waves. For a reflection hologram the interference fringes or planes are almost parallel to the emulsion surface.

Problem 3.4 Because of the angle between the object wave and the reference wave the reconstruction wave and the reconstructed object wave travel in different directions. Therefore the zeroth order of the complex holographic grating (the reconstruction wave transmitting the hologram) does not disturb the observation of the object wave. The real image wave in addition travels in a different direction compared with the virtual image wave (see Eq. (2.27)).

Problem 3.5 No, there are of course two waves, the reference wave and the object wave, scattered from the object after illumination by the reference wave. The coherence length must not be shorter than two times the distance object/emulsion.

Chapter 4

Problem 4.1 An orthoscopic image has the same curvature than the object itself. The pseudoscopic image has the opposite curvature. All convex parts of the original object are turned to be concave and vice versa. The virtual image of an object and the real image generally speaking are orthoscopic and pseudoscopic, respectively. Making holograms of holograms this maybe just vice versa.

Problem 4.2 The linear dispersion of the light is almost zero. All colors are reconstructed at the same spot. Generally a transmission hologram shows the image of the object in some distance of the hologram (see Fig. 4.2). White light reconstructs the object in all colors, which are diffracted in different directions during reconstruction (image blur). The distance between the reconstructed images of different colors increases with distance of the object from the holographic plate during exposure. If this distance is zero, the reconstructed images of all colors are at the same area. Figure 4.3 shows the reconstruction of an image plane hologram. The position of the image is at the position of the emulsion (image plane).

Problem 4.3 No. The width of the slit is about 5 mm. Diffraction angles are proportional to (λ/w), where w is the width of the slit. For λ between 400 nm and 600 nm, say 500 nm, the diffraction angle is about 0.0005 rad, which is just the aligning power of the human eye.

Problem 4.4 Full aperture holograms can be interpreted as slit aperture holograms with a very wide slit. For that reason all reconstructed images of the wide slit overlap and result in a compound color. Slit aperture holograms are exposed through a very small slit (see Problem 4.3). During white light reconstruction the slits reconstructed in different colors are separated and in distinct directions (grating theory). Therefore, the Rainbow hologram can be observed in brilliant clear colors inspecting the reconstructed image through the different slit images.

Problem 4.5 Use Appendix C to solve the problem. It is

$$F(u,v) = \int_{x=-\infty}^{x=\infty} \int_{y=-\infty}^{y=\infty} f(x,y) e^{-2\pi i(ux+vy)ux} \, dx \, dy.$$

With the given $r(x,y)$ and ξ and η instead of u and v it is

$$R(\xi,\eta) = \int_{-\infty}^{\infty} \delta(x+b,y) e^{-2\pi i(\xi x+\eta y)} \, dy.$$

This integral can be solved for $x = -b$ and $y = 0$, because of the nature of the δ function. It follows $R(\xi,\eta) = e^{-2\pi i(b\xi)} \int_{-\infty}^{\infty} \delta(0,0) \, dx = e^{-2\pi i(b\xi)}$.

Chapter 5

Problem 5.1 After Eq. (5.1) it is $\frac{1}{z_i} = \frac{z_0 z_r \pm z_c z_r \mp z_c z_0}{z_c z_0 z_r}$. Dividing all terms of the numerator by the denominator gives Eq. (5.2) for $m = 1$ and $\mu = 1$. For plane reconstruction and reference waves it is $z_c = z_r \Rightarrow \infty$. It follows $z_i = \pm z_0$. Applying Eq. (5.3) it is $0 = \frac{1}{z_{in}} + \frac{1}{z_{iu}}$. It follows that the distance of the two images is same, but on opposite sides of the hologram, which was derived before.

Problem 5.2 $V_{lat} = 1$. The answer for the last question is: Yes. If object and reference waves are the same diverging or converging waves, V_{lat} is 1, too.

Problem 5.3 Divide x_i by z_i and differentiate the result with respect to (x_0/z_0).

Problem 5.4 The radiance L of image plane holograms is given by Eq. (5.14). The pupil is the area of the H1 hologram (full aperture) and the slit, respectively. The distance of the pupils to the hologram is 15 cm. The ratio of the radiances is

$$R = \frac{L\,(\text{Rainbow})}{L\,(\text{Full aperture})} = \frac{\Omega_{p,\text{full}}}{\Omega_{p,\text{rainbow}}} = \frac{A_{\text{H1}}}{A_{\text{slit}}} = 20.$$

Problem 5.5 200 nm. From Eq. (5.15) it follows that the diffraction angle α is of the order of (λ/D); for the example given it follows $\alpha \approx 6 \times 10^{-6}$ rad, well below the resolution power of the human eye.

Chapter 6

Problem 6.1 After Eq. (6.1) it is $t(x) = \bar{t} + t_1 \cos(2\pi\sigma x)$. With the assumption that the mean transmission is 0.5 it follows: $t(x) = 0.5 + 0.5\cos(2\pi\sigma x)$. With Euler's equation it is $t(x) = 0.5 + 0.25\left(e^{i2\pi\sigma x} + e^{-i2\pi\sigma x}\right)$. In this equation the real image is $o^*(x) = 0.25\, e^{-2\pi\sigma x}$. The intensity is $I(x) = o^*(x)\,(o^*(x))^* = \frac{1}{16}$. The total intensity was 1, so the diffraction efficiency for the real image is about 6%.

Problem 6.2 After Eq. (6.4) it is $\mathbf{t}(x) = t(x)\, e^{-i\Phi(x)}$. With Eqs. (6.5) and (6.6) and small modulation it follows $\mathbf{t}(x) = e^{-i\Phi_0}\left(1 - i\Phi_1 \cos(2\pi\sigma x)\right)$. The factor $e^{-i\Phi_0}$ can be set to 1 and for the cosine function we get with Euler's function again: $\mathbf{t}(x) = 1 - \frac{\Phi_1}{2}i\left(e^{i2\pi\sigma x} + e^{-i2\pi\sigma x}\right)$. From the last equation it is obvious that the diffraction efficiency of the images is $\frac{\Phi_1^2}{4} \ll 1$. This is the maximum value for pure phase holograms.

Problem 6.3 The change of the reference wave is due to two factors: absorption and diffraction (transfer into object wave). The first part is described by αr and the second by $i\kappa o$. The same is true for the object wave. The generated object wave is absorbed to some extent by αo, and some is transferred back into the reference wave, $i\kappa r$. This is not surprising, because the object wave can be used as reconstruction wave and will reconstruct the reference wave.

Problem 6.4 The coupled equations to be solved are:

$$r'\cos\vartheta = -i\kappa o$$
$$o'\cos\vartheta = -i\kappa r \quad.$$

Differentiating the second one yields: $o''\cos\vartheta = -i\kappa r'$. Inserting the first equation into the last one results in $o''\cos\vartheta = -\frac{\kappa^2}{\cos\vartheta}o$. This differential equation is easy to solve and yields $o(z) = -i\sin\left(\frac{\pi n_1}{\lambda\cos\vartheta}z\right)$. The cosine function is a solution of the differential equation, too. With the sine function for the object wave $r(z)$ results in: $r(z) = \cos\left(\frac{\pi n_1}{\lambda \cos \vartheta} z\right)$. The sine function for the object wave and the cosine function for the reference wave fulfill the boundary conditions.

Problem 6.5 Similar to Fig. 6.4 one can draw a circle and two vectors of the same length. From the figure drawn one can derive

$\sigma = 2k \sin \vartheta$, where ϑ is the angle of $60°$.

Problem 6.6 After Eq. (6.18) it is $\varepsilon = \sin^2 \Phi$. Now it is $0.5 = \sin^2 \Phi$ and it is $\Phi = \arcsin \sqrt{(0.5)}$. Inserting all given data in Eq. (6.17) yields for $n_1 = 0.017$.

Problem 6.7 200 nm. From Eq. (5.15) it follows that the diffraction angle α is of the order of (λ/D); for the example given it follows $\alpha \approx 6 \times 10^{-6}$ rad, well below the resolution power of the human eye.

Chapter 7

Problem 7.1 Both slits emit divergent coherent light, $w(r) = \frac{A}{r} e^{i\phi}$ and $v(r) = \frac{A}{r} e^{i(\phi+\alpha)}$. ϕ is the phase, which is a function of spatial coordinates, α describes the phase difference between the both waves, and A is the amplitude. The intensity of both waves is (A^2/r^2). Coherent addition of both waves results in

$$I = (w(r) + v(r)) \cdot (w(r) + v(r))^*.$$

Here the "*" means conjugate complex. After some calculation it is $I = 2\frac{A^2}{r^2} + \frac{A^2}{r^2}\left(e^{i\alpha} + e^{-i\alpha}\right) = 2\frac{A^2}{r^2}(1 + \cos \alpha)$. The maximum value is $4(A^2/r^2)$, the minimum value is 0. The maximum is four times the value for one slit, $I_0 = (A^2/r^2)$.

Problem 7.2 After Eq. (7.3) it is $l_c = (c/\Delta f)$. For the He–Ne laser it follows $l_c = 30$ cm. For white light the bandwidth has to be calculated from the data given. Differentiating the equation $\lambda f = c$ results in $df = \frac{f}{\lambda} d\lambda$. Taking the differentials as differences and replacing f it follows $\Delta f = \frac{c}{\lambda^2}\Delta\lambda$. Inserting all data in Eq. (7.3) yields $l_c = 1.25$ μm.

Problem 7.3 To calculate the divergence angle Θ the diameter of the laser beam at the exit of the laser is neglected. Then it is $\tan\Theta = \frac{D}{2l} = \frac{0.08}{40} \approx \Theta$. After Eq. (7.5) it is $\Theta = \frac{\lambda}{\pi w_0}$ and $w_0 = 0.1$ mm.

Problem 7.4 After Eq. (7.10) it is $l_c = (2/7)$ m = 0.29 m. The bandwidth is after Eq. (7.7) $f = 1.05$ GHz.

Problem 7.5 The bandwidth of the laser is $\Delta f = (c/l_c) = 6$ GHz. In order to make a monomode laser, the distance to the next mode must be 3 GHz. With Eq. (7.11) one gets the thickness of the etalon $d = 3$ cm. The number of longitudinal modes of the original laser is $N = (2L/l_c) = 80$. Then for δf (Eq. (7.12)) one gets $\delta f = (6/80)$ GHz $= 75$ MHz, $F = 40$ and $R = 0.96$.

Problem 7.6 The minimum beam diameter w_0 can be calculated using Eq. (7.5). It is with the data given $w_0 = 0.2$ mm. Then with $s_0 = 1000$ mm and $f = 4$ mm and $y_0 = 0.2$ mm it follows $y_i = w_0' = 0.8$ µm.

Problem 7.7 After geometrical optics it is $(d/f_1) = (D/f_2)$. It follows $f_2 = 60$ cm.

Chapter 8

Problem 8.1 A half-wave plate has to be brought into the laser beam and rotated by $\pm 22.5°$ with respect to the y-direction. Inserting a polarizer into the beam with the plane of transmission at $\mp 45°$, the laser light should be extinguished.

Problem 8.2 The grating formula is $\sin \beta = N \cdot \frac{\lambda}{d_g}$. Independent of grating constant it is for the same diffraction angle $N_1 \lambda_1 = N_2 \lambda_2$. Inserting the given wavelengths it follows that the red part of the second order and the blue one of the third order overlap.

Problem 8.3 The coherence length is $2\Delta l = l_c = 30$ cm. It follows $\Delta f = \frac{c}{l_c} = 1$ GHz. The bandwidth $\Delta \lambda$ is calculated differentiating the formula $c = \lambda f$ and taking differentials as differences. Then it is $\Delta \lambda = \frac{\lambda^2}{c} \Delta f$. Inserting the given and calculated data results in $\Delta \lambda = 1.3 \times 10^{-3}$ nm. The number of longitudinal modes can be calculated roughly from Eq. (7.10) to be 5.

Problem 8.4 From the grating formula $\sin \beta = \frac{\lambda}{d_g}$ the grating constant can be derived to be $d_g = 848.7$ nm. For a He–Ne laser the angle between the two laser waves during exposure has to be $48.2°$. With the bigger mirror at $45°$ with respect to the table surface the smaller one will be at $\alpha = 24.1°$, half the angle between the two laser waves. For the green light of the Ar laser we get $\alpha = 18,6°$.

Problem 8.5 The object should be placed at the position, where the second (smaller) mirror is situated in Fig. 8.12. But this setup is not very well suited to make single beam transmission holograms. In line setups like Fig. 9.3 are used more often.

Chapter 9

Problem 9.1 In Fig. 9.1 the holographic plate and the object show the same angle with respect to the direction of wave propagation. Therefore this angle is neglected in calculation. If the reference wave intensity is set to $I_r = 1$ a.u. (arbitrary units), then the object wave is $I_o = 0.9 \times 0.7 = 0.63$. The amplitudes $A_{r,o}$ are the square roots of the intensity. It follows $A_r = 1$ and $A_o = 0.79$ a.u. The square of the sum of the amplitudes is I_{max} and the square of the difference is I_{min}. Then the visibility is $V = 0.98$.

Problem 9.2 The optical density is $D = -\log_{10} \frac{I}{I_0}$. It follows $D = -\log_{10}(1/30)$.

$D = \log_{10} 30 \approx = 1.5$.

Problem 9.3 From the second measurement (incoherent addition) it follows $I_r = 15.5$ a.u. To calculate the visibility one can use Eq. (2.11) or Eq. (2.12a). Here Eq. (2.12a) is used. It yields $V = 0.84$.

Problem 9.4 Because it is a transmission hologram, shrinkage does not affect the reconstruction of the images. It is taken $m = 1$. After Eq. (5.4b) it is for the virtual image $V_{lat} = \frac{1}{1 + \frac{z_0}{\mu z_r} - \frac{z_0}{z_r}}$. Inserting all data it follows $V_{lat} = 1.03$. For the real image the "+" sign in the denominator changes to "−". With the data given it follows $V_{lat} = 1.37$. Both images are magnified. For the virtual image the amount is very small, because of the smaller wavelength.

Chapter 10

Problem 10.1 After the first beam splitter, the ratio of illumination wave to reference wave is 4:1. The reference beam is set to $I_r = 1$ a.u. (The absolute values of the intensities are not needed to calculate the visibility, because the ratio of reference to object beam is needed only). The light scattered from the object irradiates the half sphere with a radius of 15 cm. The solid angle of the photographic plate is $\Omega = A/D^2$, A being the area of the plate and D the distance object/plate. It is $\Omega = 0.44$ rad. The full-half sphere is 2π rad. The holographic plate is 7% of the half sphere. The intensity of the object wave is $I_o = 0.07 \times 4 \times 0.7 = 0.196$ a.u.. Applying Eq. (2.12b) it follows for V: $V = 0.74$.

Problem 10.2 As calculated in Problem 10.1 the intensity of the object wave is $I_o = 0.196$ a.u. only, or less than 20% of the reference wave. Inspecting Eqs. (2.23) and (2.24) one can see that using the object wave as reconstructing wave, the amplitude of the reference wave is multiplied by the square of the amplitude of the object wave. Therefore, the reconstructed reference wave is very faint.

Problem 10.3 The maximum distance of the object from the holographic plate is the coherence length. In a single beam setup it would be just half the coherence length, because the reference wave goes through the holographic plate to illuminate the object. If the distance has to be larger, then the reflection mirror of the reference beam should be moved slightly to increase the length of the reference wave. Then, the first beam splitter has to be rotated slightly, too.

Problem 10.4 After Eq. (5.4b) it is

$$V_{lat} = \frac{m}{1 \pm \frac{m^2 \cdot z_0}{\mu \cdot z_c} - \frac{z_0}{z_c}} = \frac{1.25}{1 + \frac{1.56 \times 15}{0.79 \times 100} - \frac{15}{100}} = 1.09$$

for the virtual image. For the real image the "+" sign in the denominator changes to "−". Then it follows $V_{lat} = 2.26$. The virtual image is slightly magnified, only. If the shrinkage is not taken into account (Hariharan) then the virtual image is smaller than the object due to the shorter wavelength.

Chapter 11

Problem 11.1 As shown in Fig. 11.1 the reference wave is a plane wave. Therefore, $z_c = z_r \Rightarrow \infty$. It follows for $m = \mu = 1$ from Eq. (5.4b) $V_{lat} = 1$.

Problem 11.2 The beam splitter divides the incoming radiation by 4:1. If the reference wave is set to $I_r = 1$ a.u (a.u. = arbitrary unit), then the reconstructing wave is $I_c = 4$ a.u. The diffraction efficiency is the square of the reconstructed object wave for a reconstructing wave intensity of $I_c = 1$. Therefore, it is $o^2 = \eta \cdot I_c$. The visibility is given as $V = \frac{2\sqrt{\frac{I_r}{\eta \cdot I_c}}}{1 + \frac{I_r}{\eta \cdot I_c}}$. From the maximum value of $V = 1$ it follows $\eta = 0.25$. Exactly η should be a little bit less than 0.25 in order to result in $V < 1$.

For the given diffraction efficiency of $\eta = 0.45$ it is $o^2 = 0.45 \cdot I_c$ and $(I_r/I_c) = 0.45$. The beam splitter in Fig. 11.3 should divide the laser beam roughly by 70%:30% (exactly 69%:31%).

Problem 11.3 First the ratio (I_r/I_o) is calculated; I_r and I_o are the intensities of the reference wave and the object wave, respectively. The visibility is (see Problem 11.2) $0.80 = \frac{2\sqrt{\frac{I_r}{I_o}}}{1 + \frac{I_r}{I_o}}$. It follows $(I_r/I_o) = 4$ or $(I_r/I_o) = (1/4)$. Here the first solution is acceptable, only, because the reference wave should be always more intense than the object wave. If the diffraction efficiency is 50%, then from the incoming reconstruction wave intensity half is converted into the object wave. The reconstructing intensity should be two times the object wave intensity. Instead of 4 the ratio should be 2. Now, from the reference beam $1/4$ is used to illuminate the image area only. Therefore, the reference wave should be more intense by a factor of 4 again. The total ratio would be 8. The reflection of the beam splitter should be roughly 90% (exact value 89%). The better solution of the problem is to change the reference beam that the radiation illuminates the object area only. This shortens the exposure time.

Problem 11.4 As shown in Problem 11.3 the intensity ratio of reference and reconstruction beam should be 8. (It should be 2 because of the diffraction efficiency being 50% only. It should be 8, because the reference beam illuminates the total plate, the object covering a quarter of the plate, only.) The object beam intensity is reduced by a factor of 0.05 because of the smaller slit area compared with the original 10×10 cm^2 plate. The intensity of the reconstructing beam should be increased by a factor of 20. The final division should be 20:8 or 5:2, the reference wave intensity being fainter than the reconstructing one.

Problem 11.5 From Eq. (11.2) the focal length for H2 is calculated. It is $\frac{1}{r} + \frac{1}{s} = \frac{\mu}{f}$. With the very large distance of light source it is $s = (f/\mu)$. For the blue light the image of the slit is at $s = 30$ cm. So it follows $f = 19$ cm. Then the slit image of the green wavelength is at $s = 24$ cm and for the red it is $s = 20$ cm. The grating constant is derived from Eq. (2.35). It is with $\beta = 0$ for the blue light and $\alpha = 30°$: $d_g = 800$ nm. For the green and the red light it follows $\beta_{green} = 7°$ and $\beta_{red} = 14.5°$. With Eq. (11.2) and $\mu = 1$ the distance of the slit from H2 is 19 cm, the same as the focal length of H2. The reference wave was a plane wave. Of course, the object distance to produce H1 is 19 cm, too.

Chapter 12

Problem 12.1 The power of the reference wave is 5 mW and of the illumination wave is 20 mW. The slit transmits 6 mW within the solid angle of 2π. At the distance of the plate the power is 6 mW/r^2 = 0.027 mW/cm^2. The irradiance of the reference beam is $5/100 = 0.05$ mW/cm^2. The visibility results in $V = 0.95$. With the given sensitivity of the film the exposure time is 19.4 s. In most cases it is wise to increase this minimum value considerably.

Problem 12.2 Because of the geometry of the object, the diameter of the film cylinder must not exceed $2l_c$; two times the coherence length. Then, in reconstructing the image of the object it can be seen from the side. To see the total cone from a higher or lower position with respect to the peak of the cone, the diameter must be smaller than l_c. If the diameter is larger than $2l_c$ the cone will be invisible, starting at the peak.

Problem 12.3 The main problem is that all used lasers produce images of all colors in different directions. Therefore nine images will be reconstructed. The solution gives Fig. 12.7.

Problem 12.4 Blue laser wavelength: The grating constant is calculated to be 690 nm. The positions of the green and the red images are: green: $\alpha = 48.2°$, red: $\alpha = 66.5°$.

Green wavelength: blue $a = 42.1°$, red: $60.5°$; red wavelength: green: $a = 35°$, blue $33°$.

Because of the extension of the object the images, e.g., of the green image (blue laser) at $48°$ and the red image (green laser) at $60°$ will overlap to some extent.

Chapter 13

Problem 13.1 From Eq. (13.1a) it follows $\Delta \alpha = \frac{\ln 2}{d} = 8.6 \times 10^{-4} \mathrm{m}^{-1}$ and $0,5 = \cos \Delta \phi$ or $\Delta \phi = 1.047$. It follows (Eq. (13.1b)) $\Delta n = 0.01$.

Problem 13.2 The grating Fig. 13.2a would show many spectral orders on both sides of the zeroth order. The grating of Fig. 13.2b would show one spectrum of the first order, only. It is a volume grating.

The Fourier transform of the last one would just give the sine function itself. In case of Fig. 13.2a the transform of the stored grating being more or less a rectangular function would result in a sum of cosine functions (grating is an even function) representing different orders of spectra, observed.

Problem 13.3 As can be seen in Fig. 13.2a (right side), the minimum intensity is zero. Therefore the visibility function will give $V = \frac{I_{max}}{I_{max}} = 1$. In Fig. 13.2b the minimum intensity is $I_{min} > 0$. Therefore the visibility will give a value $V < 1$, depending on the actual value of I_{min}.

Problem 13.4 The optical depths are (Eq. (13.2) $D_{18} = 0.74$; $D_{28} = 0.55$; $D_{54} = 0.26$. The spatial frequencies are for the transparent plate $s = 180 \text{ mm}^{-1}$ and for the emulsion with 28% transmission $s = 370 \text{ mm}^{-1}$. This corresponds to grating constants of 5.5 μm and 2.7 μm, respectively.

Chapter 14

Problem 14.1 From Table 14.2 the resolution of VRM-M is 3000 mm^{-1}. The distance of Bragg planes is then $d_g = 333$ nm. The wavelength within the emulsion is $\lambda_n = 342$ nm. The maximum angle is $\delta = 61.7°$ or $118.3°$. It would be difficult but possible to make a single beam white light reflection hologram. The reference beam angle would be $28.3°$.

Problem 14.2 From Eq. (14.1) it is $d_g = \frac{\lambda_n}{2\sin\frac{\delta}{2}}$. The angle δ is $150°$, the wavelength in the material $\lambda_{He} = 422$ nm. The distance of the Bragg planes is 218.4 nm and the necessary resolution is 4577 mm^{-1}.

Problem 14.3 From the visibility equation one gets two solutions of the ratio of the reference and object wave, respectively. An additional condition is that the reference wave has to be more intense than the object wave. It is $(I_r/I_o) = 5$. From the data given one gets for the total intensity I_{total} of reference and object waves is $1.2 I_r = 2.4$ mW. The exposure time needed is at least 62.5 s (sensitivity 1500 µJ/cm^2).

Problem 14.4 After Eq. (14.3) it is $\frac{\lambda'}{\lambda} = \frac{d_{g'}}{d_g}$ and $\lambda' = 506$ nm. The hologram will show a green color.

Problem 14.5 After bleaching the emulsion shrank. The color changed from red to green. By aspirating the emulsion side the emulsion swells and the reconstruction wavelength is increased. Warming up the hologram results in shrinkage again, the amount depending from temperature and time.

Problem 14.6 The grain size of silver halide is of the order of 20–40 nm, as shown in Table 14.2. The sensitivity of dichromate gelatine is based on a molecular process with much smaller dimensions.

Problem 14.7 The reconstruction wavelength of the reflection hologram will turn to a longer wavelength because of swelling of the emulsion; the transmission hologram reconstruction wavelength will stay unchanged. The Bragg plane distance will change for the reflection hologram only.

Chapter 15

Problem 15.1 The real image is reconstructed with the conjugated complex reference wave. From Eq. (2.22a) the only interesting part is – different from Eq. (15.1) – the term $t_i(x,y) = C_2 (\mathbf{o} + \mathbf{o'})^* \cdot \mathbf{r}$. Reconstructed with $\mathbf{r}^*$ it follows $\mathbf{u}(x,y) = C_2 \cdot r^2 (\mathbf{o} + \mathbf{o'})^*$. The intensity is the complex square of the last equation, which again is Eq.(15.3).

Problem 15.2 The visibility is $V = 1$, because the minimum intensity is zero.

Problem 15.3 It is $I_e = A^2 + B^2 + AB\left(e^{i(\delta-\pi)} + e^{-i(\delta-\pi)}\right)$.
$I_e = A^2 + B^2 + 2AB\cos(\delta - \pi)$. Now it is $\cos(\delta - \pi) = -\cos(\delta)$ and $I_e = A^2 + B^2 - 2AB\cos(\delta)$. Adding and subtracting $2AB$ it follows $I_e = (A-B)^2 + 2AB(1 - \cos(\delta))$. It is $1 - \cos\delta = 2\sin^2(\delta/2)$. Inserting in the last equation follows Eq. (15.17).

Problem 15.4 The visibility is $V = \dfrac{(A+B)^2 - (A-B)^2}{(A+B)^2 + (A-B)^2} = \dfrac{2AB}{A^2+B^2}$.

Problem 15.5 From Eq. (15.25) it is $d = \dfrac{8\lambda}{2 \cdot \frac{1}{2}\sqrt{2}} = 5.6 \cdot \lambda$.

Problem 15.6 Inspecting Fig. 15.2 it is obvious that the wave has to travel the deformation path two times, similar within the Michelson interferometer. Therefore, the deformation that is the measured distance between the dark fringes corresponds to a multiple of $\lambda/2$. This is a minimum value. If α and ψ are not zero, the distance between these areas is larger and the deformation larger than $(\lambda/2)$. This is obvious, inspecting Eq. (15.25)

Problem 15.7 From Eq. (15.23) it is $\delta = \mathbf{S} \cdot \mathbf{d}$. From Fig. 15.5 it is $\delta \approx 2.4$. This is the value to produce the first minimum after Eq. (15.29). Then it follows $d \approx 0.2\lambda$. It is not possible to decide whether this is a minimum or a maximum compared with the undisturbed areas.

Chapter 16

Problem 16.1 Equation (16.1) and Fig. 16.3 are used: $1/s_{obj} + 1/s_{ref} = 1/f$. We get $1/s_{obj} = 1/f - 1/s_{ref}$. (1) We choose a parallel reference wave $s_{ref} = \infty$ and get $s_{obj} = f = 20$ cm. (2) Another choice is for example: $s_{ref} = 100$ cm and we get $s_{obj} = 25$ cm. (3) Another possibility is: $s_{ref} = 50$ cm and $s_{obj} = 33.33$ cm.

The construction has to be performed according to Fig. 16.3. It is also possible to use the same axis for the reference and object waves. Using the produced hologram as a diverging lens it has to be used according to Fig. 16.3.

Problem 16.2 Equation (16.2) and Fig. 16.3 are used: $1/s_{obj} - 1/s_{ref} = 1/f'$. We get $1/s_{obj} = 1/f' + 1/s_{ref}$. (1) We choose a parallel reference wave $s_{ref} = \infty$ and get $s_{obj} = f' = 20$ cm. (2) Another choice is for example: $s_{ref} = 100$ cm and we get $s_{obj} = 16.66$ cm. (3) Another possibility is: $s_{ref} = 50$ cm and $s_{obj} = 14.3$ cm. (4) Figure 16.5 can be used for the construction of the lens.

The construction has to be performed according to Fig. 16.3. It is also possible to use the same axis for the reference and object waves. Using the produced hologram as a collecting lens the direction of the reference beam has to be inverted for producing a real focal point and a real image.

Problem 16.3 Equation (16.1) and Fig. 16.4 are used: $1/s_{obj} + 1/s_{ref} = 1/f$. We get $1/s_{obj} = 1/f - 1/s_{ref}$. (1) We choose a parallel reference wave $s_{ref} = \infty$ and get $s_{obj} = f = 20$ cm. (2) Another choice is for example: $s_{ref} = 100$ cm and we get $s_{obj} = 25$ cm. (3) Another possibility is: $s_{ref} = 50$ cm and $s_{obj} = 33.33$ cm.

The construction has to be performed according to Fig. 16.4. It is also possible to use the same axis for the reference and object waves. Using the produced hologram as a diverging mirror it has to be used according to Fig. 16.4.

Problem 16.4 Equation (16.2) and Fig. 16.4 are used: $1/s_{obj} - 1/s_{ref} = 1/f'$. We get $1/s_{obj} = 1/s_{ref} + 1/f'$. (1) We choose a parallel reference wave $s_{ref} = \infty$ and get $s_{obj} = f' = 20$ cm. (2) Another choice is for example: $s_{ref} = 100$ cm and we get $s_{obj} = 16.66$ cm. (3) Another possibility is: $s_{ref} = 50$ cm and $s_{obj} = 14.3$ cm. (4) Figure 16.5 can be used for the construction of the mirror.

The construction has to be performed according to Fig. 16.4. It is also possible to use the same axis for the reference and object waves. Using the produced hologram as a diverging mirror the reference wave has to be inverted for producing a real focal point and a real image.

Problem 16.5 We obtain from Eq. (16.1): $1/f = 1/s_{obj} + 1/s_{ref} = (1/10 + 1/30)$ 1/cm and $f = 7.5$ cm. This focal length is obtained using the lens as a diverging lens according to Fig. 16.3. We get from Eq. (16.2): $1/f' = 1/s_{obj} - 1/s_{ref} = (1/10 - 1/30)$ 1/cm and $f' = 15$ cm. This focal length is obtained using the lens as a collecting lens. In this case the direction of the reference wave is inverted in Fig. 16.3. The reference wave comes from the opposite side (not shown in Fig. 16.3).

Problem 16.6 For a transmission grating Eq. (2.34) yields: $d_g = 1/\sigma = \lambda/(\sin\delta + \sin\delta_0) = 752$ nm. For a transmission grating the waves fall from the same side on the hologram. For a reflection hologram the directions of incidence are opposite. We get: $d_g = \lambda/(2\sin\vartheta)$ with $\vartheta + \delta/2 = 90°$ or $\vartheta = 90° - 45°/2 = 67.5°$ (see Appendix B and Section 6.3). This yields $d_g = 342$ nm.

Problem 16.7 Equation (2.35) yields: $d_g = \lambda/(\sin\alpha + \sin\beta)$. For normal incidence $\alpha = 0$ and a small value of β we get with $d_g = 500$ nm: $d_g = \lambda/\beta$ and $\beta \approx 0.126 = 7.2°$. The image size depends on the distance x from the display. If $x = 1$ m we get for the image size $a \approx \beta x = 12.6$ cm.

Chapter 17

Problem 17.1 For destructive interference the phase shift in a reflection hologram has to be $\varphi = 2\pi\Delta x/\lambda = \pi$. This yields $\Delta x = \lambda/2$.

Problem 17.2 In photoresist a surface hologram for embossing has to be created. Thus the holographic grating structure has to be rectangular to the surface plane. This is the case for a transmission hologram, which arise if the reference and object waves fall from the same direction on the holographic plate as shown in Fig. 17.1. For a reflection hologram the grating planes are more or less parallel to the surface. Thus nearly no surface relief can be created in photoresist.

Problem 17.3 Transmission holograms are created by thin phase structures on the surface (or grating structures perpendicular to the surface plane). Using a mirror or a metalized layer behind the surface transmission hologram the holographic image can be seen in reflection. Reflection holograms consist of a grating structure more or less parallel to the surface plane.

Problem 17.4 (a) The laquer layer on the surface can be removed carefully and a replica can be formed in a soft material. From this an embossing stamper can be produced. For a second method see also (b) Using the image of the hologram a new stamper can be produced.

Problem 17.5 Using a sinus grating only one diffraction order arises. Other gratings with step functions show several diffraction orders, depending on the number of steps.

Chapter 18

Problem 18.1 The complete information is stored in the holographic memory of the neuro computer. The uncomplete or incorrect information is used to readout the complete information.

Problem 18.2 Principle of phase conjugation: The reference wave 1 and the object wave 2 create a hologram in a holographic real time material (see figure). At the same time the conjugated reference wave 3 is diffracted at the hologram yielding the conjugated object wave 4.

Problem 18.3 The volume for one bit V_{bit} is given by the wavelength in the medium $V_{bit} \approx (\lambda/n)^3 = (0.4/1.5)^3 \, \mu m^3 = 0.27^3 \, \mu m^3$. The number of bits in $V = 1 \, cm^3 = 10^{12} \, \mu m$ is $N \approx V/V_{bit} = 10^{12}/0.27^3 = 0.5 \times 10^{14}$ bits.

1. Plane reference wave (to write the hologram)

2. Incoming wave (object wave)

3. Conjugated reference wave (to read the hologram)

4. Reflected wave (conjugated object wave)

Holographic real time material, e.g., photorefractive material

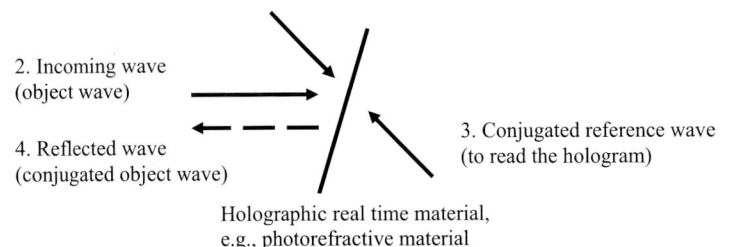

Fig. Principle of phase conjugation.

Problem 18.4 The wavelength in a medium with index of refraction is λ/n and the radius of the focal spot is given by Eq. (7.17): $w_0' \approx \lambda f/(n\pi w_0) \approx 0.23$ μm. The Rayleigh length is given by Eq. (7.16) $z_R = \pi w_0'^2 n/\lambda = 0.62$ μm. The grating period is $\lambda/2n = 0.133$ μm.

Problem 18.5 According to Fig. 7.9 the distance Δz has to be much larger than the Rayleigh length z_R: $\Delta z \gg z_R = 0.62$ μm.

Chapter 19

Problem 19.1 According to the figure below we calculate $z_1/D_1 = z_2/D_2$ or with $z = z_1 + z_2$: $z_1 = D_1 z/(D_1 + D_2)$. The angle θ can be calculated as $\tan\theta = D_1/(2z_1) = (D_1 + D_2)/(2z)$.

Hologram of the diffuser   Image of the diffuser

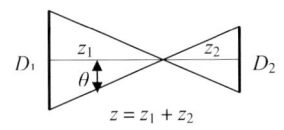

D_1 z_1 z_2 D_2
θ
$z = z_1 + z_2$

Fig. Calculation of the angular range $\pm\theta$ of a diffuser screen (according to Fig. 19.2b).

Problem 19.2 The number of pixel on the screen is $N = 1.2 \times 10^6 \cdot 2 \times 10^5 = 2.4 \times 10^{11}$. We suppose that each pixel has 10 different phases and the TV image has 30 frames per second. The number of bits per second are: $n = 2.4 \times 10^{11} \cdot 10 \times 30 \ 1/s = 7.2 \times 10^{13} \ 1/s$. This data rate is far beyond of the present technology.

Problem 19.3 It is possible to produce a hologram from two (or more) stereo images (images from different directions corresponding to the eye distance).

Problem 19.4 See Figs. 19.4 and 19.5.

Problem 19.5 Stereo images are produced using a special camera with two objectives and two films. The two images are projected on a screen with different colors or polarization. The observer has special glasses selecting one image for the right and one for the left eye.

Problem 19.6 According to Fig. 19.2a chromatic aberration cancels out over a wide range of observation angles. Aberration arise only on the outer sides of the observation field.

Chapter 20

Problem 20.1 (a) We obtain from Eq. (20.1a): $H_{\text{MPE}} = 5 \times 10^{-3}$ J/m^2. (b) It follows from Eq. (20.1b) for a large screen: $H_{\text{MPE}} = 5 \times 10^{-3} \cdot C_6$ J/m^2 = 0.33 J/m^2.

Problem 20.2 The beam radius w and the radius $R_D \geqslant w$ of the diffuser screen are given by Eq. (20.5): $R_D \geqslant w \geqslant \sqrt{\dfrac{Q \times 10^{-2} \text{ m}^2}{2\pi \cdot 0.33 \text{ J}}}$ and $R_D \geqslant w \geqslant 0.05 \cdot x$.

a) For $Q = 1$ J we get: $R_D \geqslant w \geqslant 7$ cm and for a distance of $x = 50$ cm between the diffuser and the eye: $R_D \geqslant w \geqslant 2.5$ cm.

The more severe condition gives the final result for the diameter: $2R_D \geqslant 2w \geqslant 14$ cm.

b) For $Q = 4$ J we get: $2R_D \geqslant 2w \geqslant 28$ cm.

Warning: We do not guarantee the correctness for eye savety calculations. Do not rely on these calculation. It is necessary to measure the energy density entering the eye!

Problem 20.3 According to Fig. 20.2 and Eq. (20.2) we get:

$H_0 = Q/(x^2 \pi) = 0.32$ J/m^2.

Problem 20.4 We get from Eq. (20.1a) for the maximum permissible exposure: $H_{\text{MPE}} = 5 \times 10^{-3}$ J/m^2. With $H_{\text{MPE}} \approx Q/A \approx Q/(\pi w^2)$ we get:

$Q \approx H_{\text{MPE}} \cdot \pi w^2 \approx 0.35 \times 10^{-3}$ J.

This small value for Q shows that the reference beam is really very dangerous.

Chapter 21

Problem 21.1 For radiation with normal incidence Eq. (21.1) can be used: $\sin\alpha = \lambda/d$. We get: $\alpha = 26.4°, 51.1°, 62.7°, 90°$. For $\lambda > 0.9$ µm no diffracted beam arises.

Problem 21.2 According to Eq. (16.3) we have $f' \sim 1/\lambda$ or $f_1'/f_2' = \lambda_2/\lambda_1$. For $\lambda = 0.532$ µm we get from the text above $f_{0.532}' = 40$ cm. For the other wavelength we calculate:

$$f_{0.4}' = f_{0.532}' \cdot 0.532/0.4 = 53.2 \text{ cm and } f_{0.7}' = f_{0.532}' \cdot 0.532/0.7 = 30.6 \text{ cm}.$$

A frequency doubled Nd:YAG laser is used.

Problem 21.3 According to the figure below we get: $(z+\lambda)^2 = z^2 + r^2$ or $z^2 + 2z\lambda + \lambda^2 = z^2 + r^2$. With $\lambda^2 \ll 2z\lambda$ it follows: $r \approx \sqrt{2z\lambda} = 0.355$ mm.

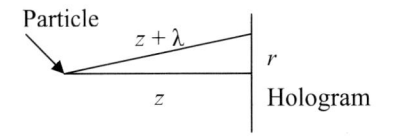

Fig. Calculation of the radius of the first interference ring in the hologram of a small particle.

Problem 21.4 The depth of field is given by Eq. (21.3): $\Delta z = D^2/\lambda$. For a particle size of $D = 1$ µm (and for $D = 10$ µm) we get $\Delta z \approx 1.6$ µm (and $\Delta z \approx 160$ µm). In holography the depth of field is much larger and given by the dimensions of the experimental arrangement (Fig. 21.7).

References

1 Bergmann, L.; Schäfer, C.; Niedrig, H. *Lehrbuch der Experimentalphysik*. Band III, Optik, Berlin: De Gruyter, 2004.

2 Collet, E. *Field Guide to Polarization*. Bellingham, WA: SPIE, 2005.

3 Hariharan, P. *Optical Holography: Principles, Techniques and Applications*. Cambridge: University Press, 2004.

4 Hecht, E. *Optics*. New York: Benjamin Cummings, 2003.

5 Atchison, D.A. *The Eye and Visual Optical Instruments*. Cambridge: Cambridge University Press, 1997.

6 Malacara, D.; Malacara, Z. *Handbook of Optical Design*. New York: Marcel Dekker, 1994.

7 Kohlrausch, F. *Praktische Physik I*. Stuttgart: Teubner, 1985.

8 Bjelkhagen, H.I. *Recording Materials for Holography and their Processing*. Berlin: Springer, 1998.

9 Kogelnik, H. *Coupled wave theory for thick hologram gratings*. Bell Syst. Techn. J. 48, 2909–2947, 1969.

10 Smith, H.M. *Holographic Recording Materials*. Berlin: Springer, 1977.

11 Denisyuk, Y.N. *Photographic reconstruction of the optical properties of an object in its own scattered radiation field*. Sov. Phys. Dokl. 7, 543–545, 1962.

12 Kogelnik, H. *Reconstruction response and efficiency of hologram gratings*. Proc. Symp. Mod. Phys., 605–617, 1967.

13 Klein, W.R.; Cook, B.D. *Unified approach to ultrasonic light diffraction*. IEEE SU 14, 123–134, 1967.

14 Eichler, J.; Eichler, H.-J. *Laser, Grundlagen, Systeme, Anwendungen*. Berlin: Springer, 2006.

15 Hariharan, P. *Basics of Holography*. Cambridge: Cambridge University Press, 2002.

16 Koechner, W. *Solid-State Laser Engineering*. Berlin: Springer, 1996.

17 Walcher, W. *Praktikum der Physik*. Stuttgart: Teubner, 2006.

18 Kneubühl, F.K.; Sigrist. M.W. *Laser*. Stuttgart: Teubner, 2005.

19 Bjelkhagen, H.I. *Silver-Halide Recording Materials*. Berlin, Heidelberg, New York: Springer, 1995.

20 Unterseher, F.; Hansen, J.; Schlesinger, B. *Holography Handbook*. Berkeley: Ross Books, 1996.

21 Bjelkhagen, H.; Canfield, H.J. *Selected papers on fundamental techniques in holography*. Bellingham, WA: SPIE, 2001.

22 Benton, S.A. *Wave front aberrations: Their effects in white light transmission holography*. Proc. Int. Symp. Disp. Hologr., 3, 167, 1985.

23 Jeong, T.H.; Sobotka, W. *Holography 2000*. Bellingham, WA: SPIE 4149, 2000.

24 *Private Communication*. Dan Schweizer invented the Rainbow-Shadowgram.

25 Jeong, T.H.; Wesly, E. *True colour holography on DuPont photopolymer material*. Holosphere 16(4), 1989.

26 Kasper, J.E.; Feller, S.A. *The Complete Book of Holograms*. New York: Dover, 2001.

27 Saxby, G. *Practical Holography*. Bristol: Institute of Physics Publishing, 2003.

28 http://www.ultimate-holography.com

29 http://www.slavich.com

Holography: A Practical Approach. Gerhard K. Ackermann and Jürgen Eichler
Copyright © 2007 WILEY-VCH Verlag GmbH & Co. KGaA, Weinheim
ISBN: 978-3-527-40663-0

30 http://www.fujihunt.com

31 http://www2.dupont.com/DuPont_Home/en_US/

32 Kreis, Th. *Handbook of Holographic Interferometry.* Weinheim: Wiley VCH, 2005.

33 Abramson, N. *Sandwich Hologram Interferometry. 5. Measurement of in Plane displacement and comparison for rigid body motion* Appl. Opt. 18, 2870, 1979.

34 Hariharan, P. *Optical Interferometry.* New York: Academic Press, 2003.

35 Abramson, N. *The holo-diagram: a practical device for making and evaluating holograms.* Appl. Opt. 8, 1235, 1969.

36 Jacquot, P. *Interferometry in Speckle light. Theory and Applications* Berlin: Springer, 2000.

37 Lee, S.H, Editor. *Selected Papers on Computer-Generated Holograms and Diffractive Optics.* SPIE Milestone Series, Bellingham: SPIE, 2006.

38 Schnars, U.; Jüptner, W. *Digital Holography.* Berlin: Springer, 2004.

39 Saxby, G. *Manual of Practical Holography*, page 162 ff, Butherworth-Heinemann, Oxford, 1991.

40 MCGrew, S. *Countermeasures against hologram counterfeiting*, http://www.nli-ltd.com.

41 van Renesse, R. L., Ed., *Optical Document Security*, ISBN0-89006-619-1, Artech House, Norwood, MA 02062, 1994

42 Hariharan, P. *Optical Holography: Principles, Techniques and Applications.* Cambridge: University Press, 1996.

43 Yu, T.S, Editor. *Optical Storage and Retrieval: Memories, Neural Networks and Fractals.* New York: Marcel Dekker, 1996.

44 Collier, R.; Burckhardt, C.; Lin, L. *Optical Holography.* Orlando: Academic Press, 1971.

45 Gaylord, T.K. *Digital Data Storage, in Handbook of Holography.* in H.J. Caulfield (Editor), London: Academics Press, 1979.

46 Savant, G.; Jannson, T.; Jannson, J. *Diffuser display screen* in J. Ludman; H. Caulfield; J. Riccobono (Editors), *Holography for the New Millenium.* page, 37ff, New York: Springer, 2002.

47 Aye, T.; Kostrzewski, A.; Savant, G.; Jannson, T., Jannson, J. *Real-time autostereoscopic 3D displays* in J. Ludman; H. Caulfield; J. Riccobono (Editors), *Holography for the New Millenium.* p. 37ff, New York: Springer, 2002.

48 Benton, S.A. *Elements of Holographic Video Imaging.* Bellingham, WA: SPIE, 1600, 82, 1991.

49 Keats, J. *The Holographic Television.* www.posci.com

50 Jacobson, A.D.; Evtulov, V. *Motion picture holography.* Appl. Phys. Letts. 14, 120, 1969.

51 Komar, V.G. *Progress of the holographic movie in the USSR.* Bellingham, WA: SPIE, 120, 127, 1977.

52 Piemontese, M. *Perspectives and renewal in 'Art and Technology'* Proc. of the Intern. Symposium on Display Holography, Bellingham, WA: SPIE, 1600, p. 166, 1991.

53 Orazem, V.; Mück, T. *Holography as a material for light – Radical holography.* Proc. of the Intern. Symposium on Display Holography, Bellingham, WA: SPIE, 1600, p. 160, 1991.

54 Benyon, M. *Art concepts in holography: works from the Male Cosmetics Series.* Proc. of the Intern. Symposium on Display Holography, Bellingham, WA: SPIE, 1600, p. 136, 1991.

55 Dawson, P. *'You are here' Landscape installation.* Proc. of the Intern. Symposium on Display Holography, Bellingham, WA: SPIE, 1600, p. 149, 1991.

56 Koechner, W. *Holographic portraiture, in Handbook of Holography.* H.J. Caulfield (Editor), New York: Academic Press, 1979.

57 *www.rotorwave.com/holoreflect.gif*.

58 Henderson, R.; Schulmeister, K. *Laser Safety.* Bristol: Institute of Physics Publishing, 2004.

59 DeFreitas, F. *Hand-made hologram portraits.* www.holoword.com/holoportrait/preface.html.

60 Riccobono, J.; Ludman, J. *Solar Holography*, in J. Ludman, H. Caulfield, J Riccobono, Editors, *Holography for the New Millenium*, Springer, New York, Berlin, Heidelberg, 2002.

61 Riccobono, J.; Ludman, J. *Solar Holography*, in H. Caulfield, Editor, The Art and Science

of Holography – A Tribute to Emmett Leith and Yuri Denisyuk, SPIE, Bellingham, 2004.

62 Pepper, A. *Architectural Holography: Building with Light, Decorating with Space*, 2002, http://www.apepper.com.

63 Thompson, B. *Particle Size Measurement*, in Handbook of Holography, pp. 609–611.

64 Kuo, C. J.; Tsai, M. H. (eds.) *Three-Dimensional Holographic Imaging*, Wiley-VCH, New York, 2002.

65 http://cmi.epfl.ch/materials/Data_S1800.pdf

66 Brown, T. G.; Creath, K.; Kogelnik, H.; Kriss, M. A.; Schmit, J.; Weber, M. J. (eds.) The Optics Encyclopedia, Wiley-VCH Weinheim, 2004.

Index

3D screen 248

a

aerosol 272
amplitude 3, 11
amplitude hologram 59
– diffraction efficiency 61
amplitude transmission 60, 169
analyzer 105
angular frequency 9
angular magnification 51
angular wave number 277
ANSI 258
architecture 271
argon laser 83
art 253
associative storage 237
autocorrelation 290

b

bandwidth 80
beam expansion 90
beam splitter 94, 107
Bessel function 214
bleaching 186, 190
bleaching bath 191
bleaching processes 187
– conventional process 187
– reversal process 187
Bragg condition 65, 282, 283
Bragg diffraction 281
Bragg reflection 32, 36
Brewster angle 92, 105, 106, 127

c

CGH, *see* computer-generated hologram
characteristic function 214
coherence 4, 75
coherence length 77, 80, 117
color holography 163
– achromatic image 167
– multilaser technique 164, 165
– rainbow technique 166
– true color hologram 164
complex amplitude 11, 279
complex number 275
computer-generated hologram 223
conjugate object wave 7
conjugated image 19
convolution 289
correlation 289
counterfeiting 235
coupled wave theory 64
– amplitude hologram 68
– phase hologram 67
coupling constant 66
cross-correlation 290

d

darkroom 121
daylighting 268
delta function 46
detection of particles 272
developer 189, 190
dichromate gelatin 192
diffraction 6, 7, 110, 112
– grating 23, 24
diffractive optical elements 223
diffuser 56
diffuser screen 246
digital micromirror device 251
diode laser 102
DMD, *see* digital micromirror device
DOE, *see* diffractive optical elements
dot matrix hologram 234
double-exposure interferometry 203
– sandwich method 205
double-sided hologram 44
DVD 241

e

eigenfrequency 97
electronic holography 224
embossed hologram 229, 232
etalon 81

Holography: A Practical Approach. Gerhard K. Ackermann and Jürgen Eichler
Copyright © 2007 WILEY-VCH Verlag GmbH & Co. KGaA, Weinheim
ISBN: 978-3-527-40663-0

Euler's relation   10
exposure   17
eye safety   258

f
finesse   81
Fourier hologram   29, 32, 46, 285
Fourier series   285
Fourier transform   46, 48, 285, 288
Fraunhofer diffraction   34
Fraunhofer hologram   29, 34
Fresnel lens   23
Fresnel zone lens   25, 30
full aperture hologram   153

g
Gabor, Dennis   7
Galilei telescope   90
gas laser   82
Gaussian beam   78, 86
geometrical optics   88
Glan–Taylor prism   93
Glan–Thompson prism   93
granulation   115
grating   112, 118, 221, 286
– reflection   119
– transmission   118

h
H1 hologram   40
H2 hologram   40
half-wave plate   93
halo   18, 19
He–Cd laser   84
He–Ne laser   82
HOE, *see* holographic optical elements
holo diagram   211
holographic diffuser screen   246
holographic display   248
holographic emulsion   169
– noise   174
– nonlinear effect   176
– optical density   169
– phase curve   169
– transmission curve   169
– visibility transfer function   172
holographic interferometry   203
– fundamental equation   209
holographic memories   240
holographic optical elements   217
holographic table   97
– compliance   100
– table top   99
hot stamping   233
Huygens, Christiaan   7

i
image aberrations   52
image equation   49
image luminance   54
– pupil   54
image processing   238
image reconstruction   131
image-plane hologram   39, 56
in-line hologram   29
index matching   111, 126, 189
interference   6, 11, 110
– fringe   12
– pattern   12
ion laser   83

k
Kepler telescope   90
Kogelnik, Herwig   65
Krypton laser   83

l
lambertian diffuser   245
laser beam   87
– beam waist   87
– divergence   113
laser mirror   96
LCD, *see* liquid crystal display
Leith, Emmett   7
Leith–Upatnieks-hologram   31
lens   107, 217
lens equation   50
light source
– spectral bandwidth   53
liquid crystal display   225, 251
longitudinal magnification   51
longitudinal mode   79

m
magnification   50
master hologram   40, 230, 261
– setup   147
master transmission hologram   256
maximum permissible exposure   258
Michelson interferometer   77, 116
mirror
– dielectric   95
– metal   95
modulation   171
– function   214
monomode operation   81
movie   251
MPE, *see* maximum permissible exposure
multiple exposure   161
multiplex hologram   162
multiplexing   261
multireflection   148

Index

n
Nd:YAG laser 84, 86, 256
neural network 239
neuro computer 238
Newton rings 111

o
object wave 3, 9
off-axis holography 16, 29, 31, 32
optical density 186
optical fiber 101
optical filtering 114
orthoscopic 19
orthoscopic image 40

p
pattern recognition 237
phase 3
phase conjugated mirror 240
phase hologram 59, 61, 185
– diffraction efficiency 62
phasor 278
photochromic material 198
photography 4
photopolymer 197
photorefractive crystals 198
photoresist 196, 230
photovoltaic concentration 265
polarization 11, 13, 91, 105
– plane 105
– prism 92, 93
polarizer 105
portrait holography 256
postswelling 189
power spectrum 176
preswelling 188
propagation constant 277
properties of the light source 52
pseudocolor 188
pseudoscopic 19
pseudoscopic image 40
pupil 55

q
q-switch 85
quater-wave plate 93

r
rainbow hologram 42
real image 19, 22, 32, 132
real-time interferometry 206
– visibility 208
reconstruction 5, 14, 18
recording 5, 14
reference wave 9
reflection hologram 29, 35, 36, 125
– duplicating methods 150
– image aberrations 151
– single-beam setup 125, 128
– split-beam setup 143
– visibility 129
– wavelength shift 133
rehalogenizing bleach bath 122
resonator length 79
ruby laser 84, 256

s
security device 234
sensitivity vector 210
shadow hologram 159
shim 233
shrinkage 188
silver halide emulsion 179, 181
– H&D curves 182
– spectral resolution 181
single beam holography 30
slit aperture hologram 153
solar energy 265
solar hologram 267
solar spectrum 266
solid-state laser 84
solving bleach bath 122
spatial coherence 57, 75
spatial filter 89, 108
spatial frequency 16, 22
speckle 56, 115, 215
– interferometry 215
– photography 215
split-beam holography 139
stereoscopic viewing 248
stereoscopy 162

t
TEM mode 79
temporal coherence 76
thermal blocking 269
thermoplastic films 194
thick hologram 31, 59
thin hologram 31, 59, 70
time average interferometry 213
transmission 17
transmission hologram 31, 121, 129
– contact copy 149
– duplicating methods 149
– ghost image 142
– single-beam setup 130
– split-beam setup 139
transversal mode 78
trouble shooting 134
TV 249

u
Upatnieks, Juris 7

v
vacuum film support 128
vibration 96
– isolating table 96
– isolator 97
virtual image 32
visibility 13
volume hologram 59, 64, 70

w
wave 9
wave equation 65
wave function 276
wave number 10
wave vector 65, 210
white light hologram 120, 152
– calculation of a rainbow hologram 155
– Dan–Schweizer hologram 160
– image plane hologram 152
– rainbow hologram 153
– rainbow hologram single-beam 160
Wiener spectrum 176
Wollaston prism 92

z
zone plate 26

Related Titles

Kreis, T.
Handbook of Holographic Interferometry
Optical and Digital Methods
554 pages with 297 figures
2005
Hardcover
ISBN: 978-3-527-40546-6

Brown, T. G., Creath, K., Kogelnik, H., Kriss, M. A., Schmit, J., Weber, M. J. (eds.)
The Optics Encyclopedia
Basic Foundations and Practical Applications. 5 Volumes
3530 pages in 5 volumes with approx. 1100 figures and approx. 100 tables
2004
Hardcover
ISBN: 978-3-527-40320-2

Kuo, C. J., Tsai, M. H. (eds.)
Three-Dimensional Holographic Imaging
224 pages
2002
Hardcover
ISBN: 978-0-471-35894-7